# ARITHMETIC
## with
## an introduction
## to
# ALGEBRA

## Martin M. Zuckerman
**CITY COLLEGE OF THE CITY UNIVERSITY OF NEW YORK**

*Ardsley House*
PUBLISHERS, INC.

**Other Titles by Martin M. Zuckerman**

*Sets and Transfinite Numbers*, Macmillan
*Intermediate Algebra: A Straightforward Approach, 2nd Edition*, John Wiley and Sons
*Elementary Algebra: A Straightforward Approach, 2nd Edition*, Allyn and Bacon
*Geometry: A Straightforward Approach, 2nd Edition*, Morton
*Arithmetic without Trumpets or Drums*, Allyn and Bacon
*Algebra and Trigonometry: A Straightforward Approach*, John Wiley and Sons
*Basic Mathematics*, Ardsley House
*Intermediate Algebra: A Straightforward Approach, Alternate Edition*, John Wiley and Sons
*Passing the City University of New York Mathematics Skills Assessment Test*, Ardsley House

Address orders and editorial
correspondence to:
Ardsley House Publishers, Inc.
A member of The Rowman & Littlefield Publishing Group
4501 Forbes Boulevard, Suite 200, Lanham, MD 20706
1-800-462-6420

ISBN 978-0-9126-7502-2

# Contents

# Preface

This book covers the basic topics in arithmetic and algebra with which every college student should be thoroughly familiar. It is written with the student in mind, in a style and at a level appropriate for student understanding.

*How to use the book:*

The format is extremely simple. Each section is divided into several topics. At the end of the section, student exercises are given that correspond to the various topics. The student should first read Topic A and work through the illustrative examples presented. Immediately following this, he or she should then attempt the exercises for Topic A at the end of the section. The student can then proceed to Topic B, followed by the exercises for Topic B, and so on.

There is a diagnostic test at the beginning of the book to help the student pinpoint strengths and weaknesses. After Part I there are two exams on arithmetic, and after Part II there are two exams on algebra. Answers to all exercises and to all exam questions are given in the back of the book. Three sample final exams are presented together with completely worked-out solutions.

*Acknowledgments:*

I gratefully acknowledge the help of a number of people who have contributed to this endeavor. I would like to thank in particular the critics for their numerous useful suggestions:

Ignacio Bello, Hillsborough Community College
Ben P. Bockstege, Broward Community College
Harold Schoen, The University of Iowa
Heschel Shapiro, Los Angeles Trade Technical College
Lora Shapiro, City College of the City University of New York
Richard Spangler, Tacoma Community College.

It was a pleasure to work with Vic Schwartz, who designed the cover, John McAusland, who did the technical illustrations, and Eileen Rosenfeld, who drew the introductions to the sections.

# Diagnostic Test

The purpose of this test is to help you understand which sections of the book you should emphasize in your studies. The question numbers correspond to the section numbers. For example, Question 1 corresponds to Section 1. If you are able to answer *all* parts of the question correctly, you may not need to study that section. But *if any part of your answer is wrong, you should study the section.* It is best to consult your instructor about your specific needs, but this test should help to pinpoint your strengths and weaknesses. When you have completed the test, turn to page 281 for the answers.

1.　**WORDS TO NUMERALS**, page　9

　　　*i.* Fifty-three thousand eight hundred eight is written _____

　　　*ii.* Fifty-eight ten-thousandths is written _____

2.　**BASIC ARITHMETIC**, page 16

　　　*i.*　$581 - 289 =$　　　　*ii.*　$597 \times 205 =$　　　　*iii.*　$10{,}302 \div 17 =$

　　　*iv.*　$905 \div 42$　　　*Answer:*　quotient =　　　　remainder =

3.　**AVERAGING**, page 23

　　　*i.* Find the average of 62, 84, 77, and 97.　　　*Answer:*

*ii.* On four chemistry exams, Judy's grades are 85, 65, 0, and 90. The average of her grades is —————————

## 4. ADDING AND SUBTRACTING UNITS, page 27

*i.* A flight left at 7:52 A.M. and arrived at 11:25 A.M. that morning. How long did it last?  *Answer:*

*ii.*    9 pounds  9 ounces
       −4 pounds 10 ounces
       ———————————————

*iii.*    4 feet 7 inches
        +3 feet 9 inches
        ———————————————

## 5. ADDING AND SUBTRACTING FRACTIONS, page 33

*i.* $\frac{1}{5} + \frac{3}{4} =$

*ii.* $\frac{5}{6} - \frac{1}{8} =$

## 6. MULTIPLYING AND DIVIDING FRACTIONS, page 43

*i.* $\frac{5}{8} \times \frac{2}{15} =$

*ii.* $\frac{3}{8} \div \frac{9}{4} =$

## 7. MIXED NUMBERS, page 49

*i.* $2\frac{2}{3} + 1\frac{1}{6} =$

*ii.* $2\frac{1}{2} \times 3\frac{1}{4} =$

*iii.* $6\frac{1}{4} \div 3\frac{1}{8} =$

## 8. SIZE OF FRACTIONS, page 57

*i.* Which of these fractions is the smallest?

(A) $\frac{2}{5}$     (B) $\frac{1}{3}$     (C) $\frac{3}{8}$     (D) $\frac{2}{7}$     (E) $\frac{3}{10}$

## 9. SIZE OF DECIMALS, page 64

*i.* Which number is the smallest?

(A) .404     (B) .411     (C) .4009     (D) .4010     (E) .4001

## 10. ADDING AND SUBTRACTING DECIMALS, page 70

*i.* $5.905 + 7.08 + 12 =$

*ii.* $29.6 - 19.79 =$

11.  **MULTIPLYING AND DIVIDING DECIMALS,** page 73

    *i.* Multiply.    4.4 × .06

    *ii.* A gallon of paint costs $9.95. What is the cost of 39 gallons of that paint?
    *Answer:*

    *iii.* 720 ÷ .12 =

    *iv.* Cherries sell for $1.40 per pound. How many pounds can be bought for $11.20?
    *Answer:*

12.  **COST AND PROFIT,** page 78

    *i.* Find the total cost of 6 bottles of detergent at $2.25 per bottle and 8 rolls of paper towels at $1.04 per roll.  *Answer:*

    *ii.* At a concert, 632 tickets are sold at $7.50 each. It costs $1800 to rent the hall and $500 for other expenses. The profit is _____

    *iii.* A limousine service charges $25 for the first hour and $15 for each additional hour. The charge for 6 hours of limousine service is _____

13.  **FRACTIONS, DECIMALS, AND PERCENT,** page 82

    *i.* Change $\frac{2}{11}$ to a decimal rounded to the nearest hundredth.
    *Answer:*

    *ii.* Change .012 to a fraction in lowest terms.  *Answer:*

    *iii.* What is 56% expressed as a fraction in lowest terms?  *Answer:*

14.  **PERCENT PROBLEMS,** page 88

    *i.* What is 40% of 60?  *Answer:*

    *ii.* If 20% of a number is 15, what is the number?  *Answer:*

    *iii.* A pitcher throws 75 called strikes out of a total of 125 pitches in a game. What percent of his pitches are called strikes?
    *Answer:*

    *iv.* An alloy contains 40% copper. How much copper is there in 250 tons of the alloy?
    *Answer:*

15.  **PERCENT INCREASE OR DECREASE,** page 96

    *i.* The price of gasoline, which was $1.20 per gallon, is increased by 5%. The new price per gallon is _____

    *ii.* A clock that was selling for $25 was reduced by 25%. The new price is _____

## 16. AREA AND PERIMETER, page 101

*i.* How much does it cost to carpet a room that is 18 feet by 12 feet, if the carpeting costs $3 per square foot? *Answer:*

*ii.* A woman wants to sew a silk border around a blanket that is 8 feet by 5 feet. How long a piece of silk does she need? *Answer:*

*iii.* The area of a rectangle is 100 square inches and the length is 20 inches. Find the perimeter of the rectangle. *Answer:*

## 17. GRAPHS, page 111

*i.* The graph at the right shows the number of students who graduated from Hoover High School each year from 1978 to 1982. The total number that graduated during the years 1978, 1979, and 1980 was closest to

(A) 10,000        (B) 11,000

(C) 11,500        (D) 12,000

(E) 13,000

Graduates of
Hoover High School

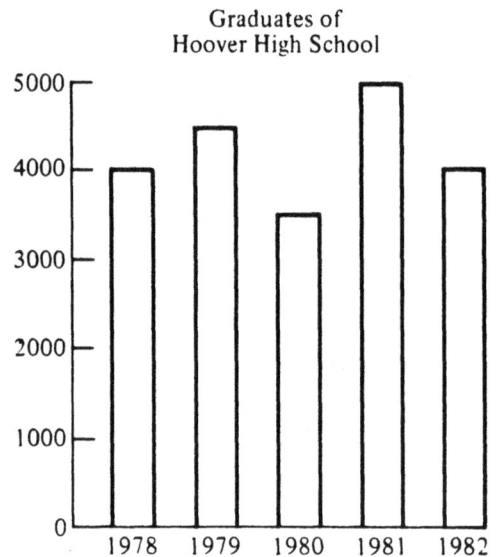

*ii.* The graph at the left, indicates a family's allotment of their yearly income of $25,000. For which item did the family spend $5000?

(A) food            (B) services

(C) clothing        (D) taxes

(E) rent

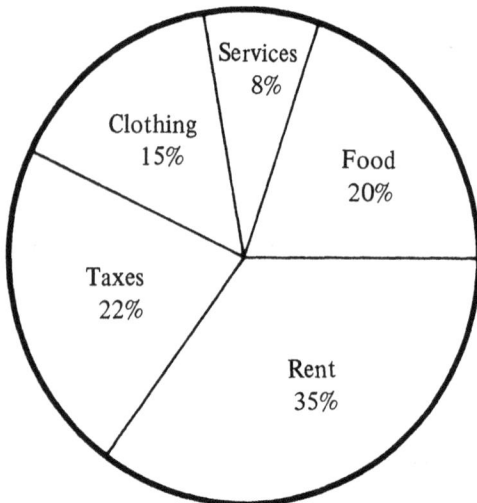

Yearly Income: $25,000

*iii.* The graph at the right indicates the annual profits of a textile manufacturer, in thousands of dollars. What was the decrease in profits from 1978 to 1979?

Corporate Profits

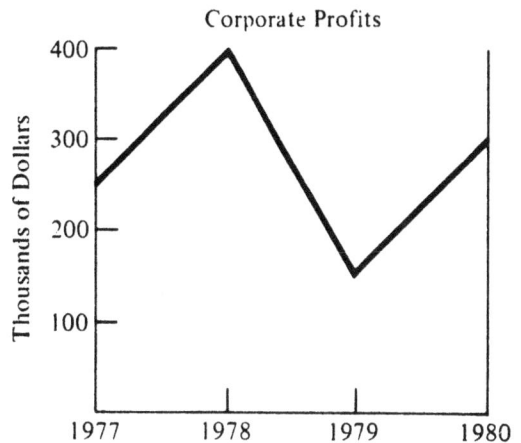

(A) $150      (B) $250

(C) $150,000      (D) $250,000

(E) $400,000

18. ARITHMETIC WITH NEGATIVE INTEGERS, page 119

    *i.*   $10 - (-3) =$        *ii.*   $(-5)(-6) =$        *iii.*   $\dfrac{27}{-9} =$

19. SEVERAL OPERATIONS, page 129

    *i.*   $4(2 + 3) - 5^2 =$        *ii.*   $(-3)(-2)^2 - 5(-1)^2 =$

20. ALGEBRAIC EXPRESSIONS, page 139

    *i.* If $x$ represents a number, express 2 more than 3 times the number.  *Answer:*

    *ii.* Find the value, in dollars, of $x$ five-dollar bills and $y$ ten-dollar bills. *Answer:*

21. SUBSTITUTING NUMBERS FOR VARIABLES, page 145.

    *i.* Find the value of $2x^2 + 5x$ when $x = 4$.  *Answer:*

    *ii.* Find the value of $4p^2 - 3pq$ when $p = -2$ and $q = 3$.  *Answer:*

22. ADDING AND SUBTRACTING POLYNOMIALS, page 153

    *i.* Add $4x + 5y$ and $7x - 4y$.    *Answer:*

    *ii.* Subtract $6x - 5$ from $8x^2 + 3$.   *Answer:*

23. MULTIPLYING POLYNOMIALS, page 160

    *i.* Multiply $5x^2y^3$ and $-4x^4y$.  *Answer:*

    *ii.* $4p^2(pq - 2q^2) =$

24. POWERS OF MONOMIALS, page 168

    *i.* Simplify $4x^2(5x^2y^4)^2$.  *Answer:*

### 25.  DIVIDING POLYNOMIALS, page 173

i.  $\dfrac{20a^2 + 24a}{4a} =$

### 26.  COMMON FACTORS, page 178

i.  Factor $15a^3b^2 + 25a^2b$.    *Answer:*

### 27.  SOLVING EQUATIONS, page 185

i.  Solve for $x$:

$$5 - 3x = 2x - 5$$

*Answer:*

### 28.  EQUATIONS WITH FRACTIONS, page 193

i.  Find $t$ if $\dfrac{t-3}{9} = \dfrac{t-5}{3}$.    *Answer:*

ii.  If $\dfrac{y}{2} - 6 = \dfrac{y}{5}$, then $y =$

### 29.  LITERAL EQUATIONS, page 199

i.  If $5a + b - 2c = 10$, then $c =$

ii.  If $\dfrac{2+3x}{5} = y$, then $x =$

### 30.  x, y-PLANE, page 207

i.  For each point in the figure at the right, find:

(a)  its $x$-coordinate,

(b)  its $y$-coordinate.

(c)  Express the point in terms of its coordinates.

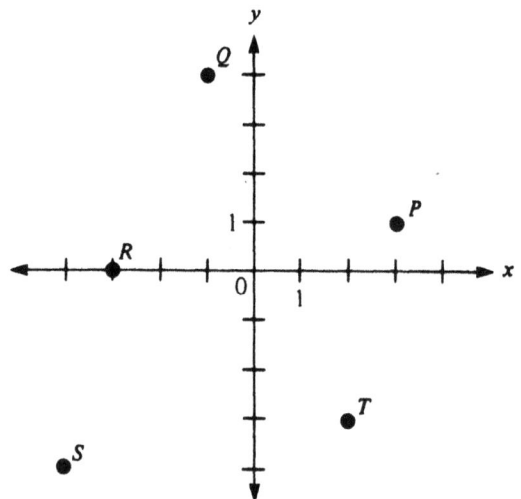

| | | | |
|---|---|---|---|
| P: | (a) | (b) | (c) P = |
| Q: | (a) | (b) | (c) Q = |
| R: | (a) | (b) | (c) R = |
| S: | (a) | (b) | (c) S = |
| T: | (a) | (b) | (c) T = |

**31. GRAPH OF A LINE, page** 213

   *i.* Which of these points lies on the graph of $2x + y = 12$?

      **(A)** $(2, 1)$      **(B)** $(3, 5)$      **(C)** $(1, 8)$      **(D)** $(4, 4)$      **(E)** $(5, 4)$

   *ii.* Which of these lines is horizontal?

      **(A)** $x + y = 6$    **(B)** $x = 6$      **(C)** $y = 6$      **(D)** $x - y = 6$     **(E)** $y = 6x$

**32. EQUATION OF A LINE, page** 222

   *i.* At what point does the graph of $y = -3 - 2x$ intersect the $y$-axis?   *Answer:*

   *ii.* If the point $(0, b)$ lies on the graph of $2y = 3x - 4$, then $b =$

   *iii.* An equation for the line at the right is

      **(A)** $y = 4 - x$         **(B)** $y = 4 - 2x$

      **(C)** $y = 4$             **(D)** $x = 4$

      **(E)** $x + 2y = 4$

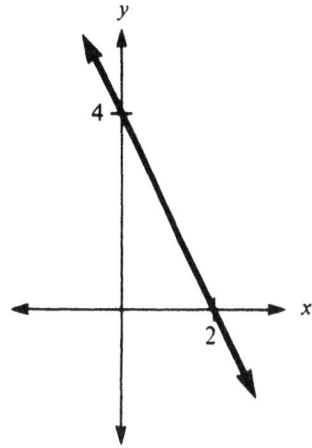

**33. SYSTEMS OF EQUATIONS, page** 233

   *i.* If $3x + 4y = 7$ and $5x + 2y = 7$, then

      **(A)** $x = 2, y = 0$             **(B)** $x = 1, y = 1$             **(C)** $x = 5, y = -2$

      **(D)** $x = -3, y = 4$         **(E)** $x = 9, y = -5$

   *ii.* If $y = 5x - 1$ and $3x + 4y = -4$, then

      **(A)** $x = 1, y = 4$            **(B)** $x = 2, y = 9$            **(C)** $x = 0, y = -1$

      **(D)** $x = -1, y = -6$      **(E)** $x = \dfrac{1}{5}, y = 0$

**34. APPLIED PROBLEMS, page** 243

   *i.* It costs a manufacturer $2 to produce each wallet. In addition, there is a general overhead cost of $5000. He decides to produce 8000 wallets. If he receives $10 per wallet from a wholesaler, how many wallets must he sell to make a profit of $10,000?   *Answer:*

    *ii.* Three out of every 5 cars sold one month have radial tires. If an agency sells 15 cars with radial tires that month, how many cars do they sell that month? *Answer:*

    *iii.* Twenty percent of the 400 employees in a factory are women. In order to have 50% women employees in the factory, how many additional women must be hired? *Answer:*

**35.** SQUARE ROOTS AND RIGHT TRIANGLES, page 251

    *i.* $\sqrt{8^2 - 5^2} =$

    *ii.* In the right triangle shown here, find $x$.

       *Answer:*

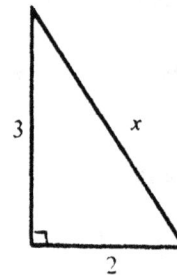

# Part One
# Arithmetic

---

## 1. Words to Numerals

Sometimes a number is written in words. You must then write it in numeral form, that is, in terms of the **digits**.

> 0, 1, 2, 3, 4, 5, 6, 7, 8, 9

For example,

> eight hundred fifty-six    is written    856

as a 3-digit number.

### A. INTEGERS

Note the correspondence:

| | |
|---:|:---|
| 1's | ←——→ ones |
| 10's | ←——→ tens |
| 100's | ←——→ hundreds |
| 1000's | ←——→ thousands |
| 10,000's | ←——→ ten-thousands |
| 100,000's | ←——→ hundred-thousands |
| 1,000,000's | ←——→ millions |

When writing integers with *more than four digits*, begin at the right and use commas to separate the digits into groups of three. For example,

```
millions
    thousands
12,607,259
```

This represents

　　12 million, 607 thousand, 259

**Example 1 ▶**  Write "seven thousand four hundred fifty-two" in numerals.

***Solution.***

```
1000's
      100's
           10's
               1's
  7    4    5   2
```

Thus write 7452.　　　　　　　　　　　　　　　　　　　　　　　◀

**Example 2 ▶**  Write "forty thousand two hundred sixty" in numerals.

***Solution.***

```
10,000's
       1000's
             100's
                  10's
                      1's
  4    0,   2    6    0
```

Thus write 40,260.　　　　　　　　　　　　　　　　　　　　　◀

**Example 3 ▶**  Write "eight hundred thousand four hundred" in numerals.

***Solution.***

```
100,000's
        10,000's
               1000's
                     100's
                          10's
                              1's
  8    0    0,   4    0    0
```

Thus write 800,400.　　　　　　　　　　　　　　　　　　　　◀

**Example 4 ▶**   Write "nine million twelve thousand seventy-eight" in numerals.

*Solution.*

```
1,000,000's
  |   100,000's
  |     |   10,000's .
  |     |     |   1000's .
  |     |     |     |   100's
  |     |     |     |     |   10's
  |     |     |     |     |     |   1's
  ↓     ↓     ↓     ↓     ↓     ↓   ↓
  9,    0     1     2,    0     7   8
```

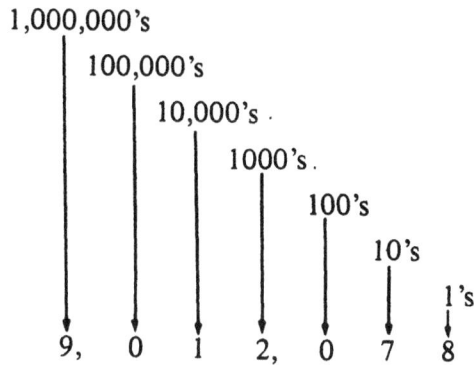

Thus write 9,012,078.                                                    ◀

*Try the exercises for Topic A on page 12.*

## B. DECIMALS

Digits to the *right* of the *decimal point* are called **decimal digits**. The first decimal digit indicates the number of *tenths*; the second decimal digit, the number of *hundredths*; the third decimal digit, the number of *thousandths*; the fourth decimal digit, the number of *ten-thousandths*, and so on.

**Example 5 ▶**   Write "three thousand five hundred eight ten-thousandths" in numerals.

*Solution.*

```
        .3  5  0  8
         ↑  ↑  ↑  ↑
   tenths   |  |  |
   hundredths  |  |
   thousandths    |
   ten-thousandths
```
◀

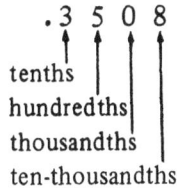

**Example 6 ▶**   Write  (a) seven tenths   (b) seven hundredths   (c) seven thousandths
in numerals.

*Solution.*

(a)     .7              (b)     .0  7            (c)     .0  0  7   ◀
         ↑                       ↑  ↑                     ↑  ↑  ↑
      tenths                  tenths |                 tenths |  |
                               hundredths            hundredths |
                                                     thousandths

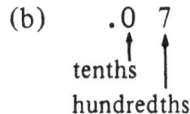

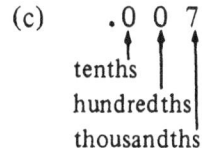

When expressing 466 in words we say

four hundred sixty-six

and *not* "four hundred and sixty-six". The word "and" is used when a number has both an integral part and a decimal part, as in Example 7.

**Example 7 ▶**   Write "four hundred and sixty-six thousandths" in numerals.

*Solution.*

```
100's
  │   10's
  │    │    1's
  │    │    │
  ↓    ↓    ↓
  4    0    0  .  0    6    6             ◀
                     ↑    ↑    ↑
                   tenths │    │
                   hundredths│
                   thousandths
```

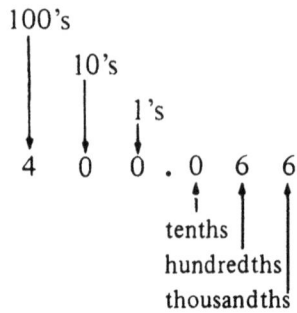

        Note the difference between 10's (tens) and ten*ths*, between 100's (hundreds) and hundred*ths*, and so on.

*Try the exercises for Topic B on page 13.*

## C.  MULTIPLE CHOICE

**Example 8 ▶**   "Three hundred thousand five hundred" is written

(A)  3500                        (B)  305,000                        (C)  300.05

(D)  300,500                     (E)  300,000.05

*Solution.*

```
100,000's
    │    10,000's
    │        │    1000's
    │        │        │    100's
    │        │        │        │    10's
    │        │        │        │        │    1's
    ↓        ↓        ↓        ↓        ↓    ↓
    3        0        0   ,    5        0    0
```

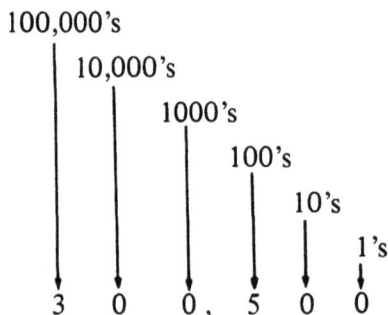

The correct choice is (D).                                                  ◀

*Try the exercises for Topic C on page 14.*

## EXERCISES

    *Answers to all exercises are given, beginning on page 281.*

A.  *Write in numerals.*

   1.  two hundred forty-five           *Answer:*

   2.  six hundred nine          *Answer:*

   3.  seven hundred eighty            *Answer:*

 4.  one thousand two hundred thirty-six          *Answer:*

 5.  seven thousand six hundred two          *Answer:*

 6.  eight thousand forty-four          *Answer:*

 7.  six thousand five hundred          *Answer:*

 8.  twenty-four thousand five hundred          *Answer:*

 9.  sixty-three thousand fifty-six          *Answer:*

10.  thirty-seven thousand nine          *Answer:*

11.  four hundred thousand two hundred twelve          *Answer:*

12.  three hundred fifty thousand six hundred          *Answer:*

13.  seven hundred forty-seven thousand eight          *Answer:*

14.  six hundred four thousand five hundred thirty          *Answer:*

15.  one hundred ninety-eight thousand two hundred forty          *Answer:*

16.  six million thirty-five thousand          *Answer:*

17.  two million one thousand ninety-six          *Answer:*

18.  five million nineteen thousand four hundred fifty          *Answer:*

19.  ten million one hundred sixty-three thousand          *Answer:*

20.  ninety million twenty thousand fourteen          *Answer:*

**B.**  *Write in numerals.*

21.  four tenths          *Answer:*

22.  fifty-eight hundredths          *Answer:*

23.  one hundred thirty-six thousandths          *Answer:*

24.  five thousand four hundred twenty-seven ten-thousandths          *Answer:*

25.  three tenths          *Answer:*

26.  three hundredths          *Answer:*

27.  three thousandths          *Answer:*

28.  thirty-eight hundredths          *Answer:*

29.  thirty-eight thousandths          *Answer:*

30.  three hundred eight thousandths          *Answer:*

31.  three hundred eight ten-thousandths          *Answer:*

32.  three hundred eight thousand          *Answer:*

33.  forty-nine and six tenths          *Answer:*

34.  five hundred nine and seven hundredths          *Answer:*

35.  six thousand one hundred and five tenths          *Answer:*

36.  eight hundred thirteen and fifteen thousandths          *Answer:*

C.  *Draw a circle around the correct letter, as illustrated in Exercise 37.*

37.  Five hundred eight is written

(A) 508          (B) 5008          (C) 580          (D) .508          (E) 500.8

38.  Forty-two thousand six hundred is written

(A) 426          (B) 4260          (C) 42,600          (D) 42,006          (E) 42,000.06

39.  Five million eighty-four thousand five is written

(A) 5,840,500          (B) 5,840,005          (C) 5,084,500

(D) 5,084,005          (E) 5,084,050

40.  Thirty-eight hundredths is written

(A) 3800          (B) 38,000          (C) 038          (D) .38          (E) .038

41.  Five hundred five thousandths is written

(A) 505          (B) 505,000          (C) .505          (D) .0505          (E) 500.005

42.  Three hundred and seven tenths is written

(A) 307          (B) 300.7          (C) 307.7          (D) 300.07          (E) 307.07

43.  Nine million eight hundred thousand is written

(A) 9,000,800          (B) 9,800,000          (C) 9,800,008

(D) 9,000,000.008          (E) 9,000,000.08

44.  Three hundred thousand six is written

(A) 300,060          (B) 300,006          (C) 300,000.6

(D) 300,000.06          (E) 306,000

45.  Six hundred and six thousandths is written

(A) 606,000          (B) 600,000.006          (C) 600,600

(D) 600.06          (E) 600.006

46. Seven hundred fifty thousand four hundred is written

(A) 750.04          (B) 754.04          (C) 750,000.04

(D) 750,400          (E) 754,000

47. Three hundred five thousandths is written

(A) 305,000          (B) 300,005          (C) .0305

(D) 300,000.005          (E) .305

48. Nine and five hundredths is written

(A) 9.5          (B) 9.05          (C) 9.005          (D) 9500          (E) 509

49. Sixty-three ten-thousandths is written

(A) 63,000          (B) 63,010          (C) 63.010

(D) .063          (E) .0063

50. Four million eighteen is written

(A) 4,000,018          (B) 4,018,000          (C) 4,000,000.18

(D) 4,000,000.018          (E) 4,180,000

51. Five hundred fifty-five thousand four hundred ninety-three is written

(A) 550,493          (B) 555,493          (C) 555.493

(D) 555,000.493          (E) .555 493

52. Six thousand four and five hundredths is written

(A) 604.05          (B) 6004.05          (C) 6004.005

(D) 604,500          (E) 6,004,500

# 2.  Basic Arithmetic

Let us briefly review the basic processes of adding, subtracting, multiplying and dividing integers.

## A. ADDITION

DEFINITION

> When two or more numbers are added, the result is called the **sum**.

**Example 1** ▶  $276 + 798 =$

*Solution.* Arrange in columns and carry the 1's as indicated.

| *Step* 1. | | | | *Step* 2. | | | | *Step* 3. | | |
|---|---|---|---|---|---|---|---|---|---|---|
| 100's | 10's | 1's | | 100's | 10's | 1's | | 100's | 10's | 1's |
| | 1 | | | 1 | 1 | | | 1 | 1 | |
| 2 | 7 | 6 | | 2 | 7 | 6 | | 2 | 7 | 6 |
| + 7 | 9 | 8 | | + 7 | 9 | 8 | | + 7 | 9 | 8 |
| | | 4 | | | 7 | 4 | | 10 | 7 | 4 |

The sum is 1074.                                                              ◀

*Try the exercises for Topic A on page 20.*

## B. SUBTRACTION

DEFINITION

> When one number is subtracted from another, the result is called the **difference**.

There are two methods of subtraction that are commonly used. Both are illustrated.

**Example 2** ▶    $315 - 178 =$

*Solution.*

*Step* 1. Borrow from the 10's column in order to increase 5 to 15 in the 1's column.

$$
\begin{array}{ccc}
\begin{array}{r}
0\ \boxed{15}\\
3\cancel{1}\ \boxed{\cancel{5}}\\
-\ 17\ \ 8\\
\hline
7
\end{array}
&
\text{or}
&
\begin{array}{r}
\boxed{15}\\
31\ \boxed{\cancel{5}}\\
8\\
-\ 1\cancel{7}\ \ 8\\
\hline
7
\end{array}
\end{array}
$$

*Step* 2. Borrow 10 tens from the 100's column.

$$
\begin{array}{ccc}
\begin{array}{r}
\boxed{10}\\
2\ \ \cancel{0}\\
\cancel{3}\ \cancel{1}\ \ 5\\
-\ 1\ \ 7\ \ 8\\
\hline
3\ \ \ 7
\end{array}
&
\text{or}
&
\begin{array}{r}
\boxed{11}\\
3\ \cancel{1}\ \ 5\\
2\ \ 8\\
-\ \cancel{1}\ \cancel{7}\ \ 8\\
\hline
3\ \ 7
\end{array}
\end{array}
$$

*Step* 3.

$$
\begin{array}{ccc}
\begin{array}{r}
2\\
\cancel{3}\ \ 15\\
-\ 1\ \ 78\\
\hline
1\ \ 37
\end{array}
&
\text{or}
&
\begin{array}{r}
3\ \ 15\\
2\\
-\ \cancel{1}\ \ 78\\
\hline
1\ \ 37
\end{array}
\end{array}
$$

The difference is 137.    ◀

**Example 3** ▶    $504 - 169 =$

*Solution.* Write

$$
\begin{array}{r}
50\ \boxed{4}\\
-\ 16\ \boxed{9}
\end{array}
$$

You want to borrow from the 10's column in order to subtract

$$
\begin{array}{r}
14\\
-\ 9\\
\hline
\end{array}
$$

in the 1's column. Because 0 is the 10's digit of 504, you must first borrow from the 100's column.

$$
\begin{array}{ccc}
\begin{array}{r}
4\ \ 10\\
\cancel{5}\ \cancel{0}\ \ 4\\
-\ 1\ \ 6\ \ 9\\
\hline
\end{array}
&
\text{or}
&
\begin{array}{r}
10\\
5\ \cancel{0}\ \ 4\\
2\\
-\ \cancel{1}\ 6\ \ 9\\
\hline
\end{array}
\end{array}
$$

Now proceed as before.

$$
\begin{array}{r}
9 \\
4\ \cancel{10}\ 14 \\
\cancel{5}\ \cancel{0}\ 4 \\
-1\ 6\ 9 \\
\hline
3\ 3\ 5
\end{array}
\qquad \text{or} \qquad
\begin{array}{r}
10\ \ 14 \\
5\ \cancel{0}\ \cancel{4} \\
2\ \ 7 \\
-\cancel{1}\ \cancel{6}\ 9 \\
\hline
3\ \ 3\ \ 5
\end{array}
$$

The difference is 335.                                                          ◀

*Try the exercises for Topic B on page 21.*

## C. MULTIPLICATION

DEFINITION

> When two or more numbers are multiplied, each is called a **factor**, or a **divisor**, of the resulting **product**. For example, 2 and 3 are each factors (divisors) of 6 because the product of 2 and 3, or 2 × 3, is 6. Also, 6 is called a **multiple** of 2 and 3.

**Example 4** ▶  485 × 329 =

*Solution.* Write in column form:

$$
\begin{array}{r}
4\ 8\ 5 \\
\times\ 3\ 2\ 9 \\
\hline
4\ 3\ 6\ 5 \\
9\ 7\ 0\ \boxed{0} \\
1\ 4\ 5\ 5\ \boxed{0}\ \boxed{0} \\
\hline
1\ 5\ 9\ 5\ 6\ 5
\end{array}
$$

- - - - - - 485 × 9
- - - - - - 485 × 2$\boxed{0}$ = 970$\boxed{0}$
- - - - - - 485 × 3$\boxed{00}$ = 1455$\boxed{00}$

The product is 159,565.                                                          ◀

**Example 5** ▶  7831 × 5080 =

*Solution.* When one or more digits of the second factor, here 5080, is 0, the method can be shortened.

$$
\begin{array}{r}
7\ 8\ 3\ 1 \\
\times\ 5\ 0\ 8\ 0 \\
\hline
6\ 2\ 6\ 4\ 8\ \boxed{0} \\
3\ 9\ 1\ 5\ 5\ \boxed{0}\ 0\ 0 \\
\hline
3\ 9\ 7\ 8\ 1\ 4\ 8\ 0
\end{array}
$$

- - - - - - 7831 × 8$\boxed{0}$ = 62648$\boxed{0}$
- - - - - - 7831 × 5$\boxed{0}$00 = 39155$\boxed{0}$00

The product is 39,781,480.                                                          ◀

*Try the exercises for Topic C on page 21.*

## D. DIVISION

DEFINITION

> Clearly 6 ÷ 2 = 3. Here 6 is called the **dividend**, 2 the **divisor**, and 3 the **quotient**.

**Example 6 ▶**     $4902 \div 86 =$

*Solution.*  Here 4902 is the dividend and 86 is the divisor. Write

$$86 \overline{)4902}$$

The divisor, 86, is a 2-digit number. Now consider $\overline{49}02$, whose first two digits are $\boxed{49}$. Because

$$86 > 49 \qquad (86 \text{ is greater than } 49)$$

you must consider the first three digits $\boxed{490}$ and divide:

$$\frac{\boxed{49}0}{\boxed{8}6}$$

Because $\dfrac{49}{8} > 6$, *try* 6.

$$\begin{array}{r} 6 \\ 86 \overline{)4902} \\ 516 \end{array} \qquad \text{Too large!}$$

Next try 5.

$$\begin{array}{r} 5 \\ 86 \overline{)4902} \\ 430 \\ \hline 60 \end{array}$$

Bring down the next digit of the divisor.

$$\begin{array}{r} 5 \\ 86 \overline{)4902} \\ 430 \\ \hline 602 \end{array}$$

Now divide 602 by the divisor, 86.

$$\frac{\boxed{60}2}{\boxed{8}6}$$

Because $\dfrac{60}{8} > 7$, *try* 7.

$$\begin{array}{r} 57 \\ 86 \overline{)4902} \\ 430 \\ \hline 602 \\ 602 \end{array}$$

The quotient is 57.                                    ◀

*Try the exercises for Topic D on page 21.*

## E.  DIVISION WITH A REMAINDER

Sometimes when you divide, there is a *remainder*. For example,

$$\begin{array}{r} 3 \\ 21\overline{)67} \\ 63 \\ \hline 4 \end{array}$$

We say that 3 is the **quotient** and that 4 is the **remainder**. The remainder is always smaller than the divisor.

**Example 7 ▶** Find (a) the quotient and (b) the remainder.

3888 ÷ 19

*Solution.*

$$\begin{array}{r} 204 \\ 19\overline{)3888} \\ 38 \\ \hline 88 \\ 76 \\ \hline 12 \end{array}$$

Observe that 12 is smaller than 19, and there are no more digits in the dividend to bring down.

(a)  The quotient is 204.        (b)  The remainder is 12.        ◀

*Try the exercises for Topic E on page 21.*

## F.  APPLICATIONS

**Example 8 ▶** A sum of four thousand eight hundred dollars is divided equally among six business partners. How much is each partner's share?

*Solution.*  Divide 4800 by 6.

4800 ÷ 6 = 800

Each partner's share is $800.        ◀

*Try the exercises for Topic F on page 22.*

## EXERCISES

**A.** *Add.*

1.  432 + 526        *Answer:*        2.  818 + 362        *Answer:*

3.  785 + 860        *Answer:*        4.  799 + 894        *Answer:*

**5.** 1065 + 595    *Answer:*        **6.** 8294 + 4493    *Answer:*

**7.** 296 + 704    *Answer:*        **8.** 8096 + 219    *Answer:*

**B.** *Subtract.*

**9.** 285 – 62    *Answer:*        **10.** 329 – 92    *Answer:*

**11.** 714 – 47    *Answer:*        **12.** 108 – 49    *Answer:*

**13.** 594 – 263    *Answer:*        **14.** 417 – 208    *Answer:*

**15.** 362 – 253    *Answer:*        **16.** 743 – 248    *Answer:*

**17.** 881 – 197    *Answer:*        **18.** 4020 – 3018    *Answer:*

**19.** 1989 – 998    *Answer:*        **20.** 5084 – 3989    *Answer:*

**21.** 594 – 388    *Answer:*        **22.** 883 – 466    *Answer:*

**C.** *Multiply.*

**23.** 522 × 43    *Answer:*        **24.** 207 × 81    *Answer:*

**25.** 345 × 213    *Answer:*        **26.** 486 × 306    *Answer:*

**27.** 786 × 590    *Answer:*        **28.** 872 × 565    *Answer:*

**29.** 8043 × 621    *Answer:*        **30.** 7205 × 106    *Answer:*

**31.** 6043 × 3908    *Answer:*        **32.** 4925 × 3479    *Answer:*

**33.** 384 × 404    *Answer:*        **34.** 3019 × 192    *Answer:*

**D.** *Divide.*

**35.** 713 ÷ 23    *Answer:*        **36.** 1428 ÷ 42    *Answer:*

**37.** 936 ÷ 36    *Answer:*        **38.** 893 ÷ 47    *Answer:*

**39.** 1024 ÷ 64    *Answer:*        **40.** 2592 ÷ 48    *Answer:*

**41.** 8060 ÷ 26    *Answer:*        **42.** 1530 ÷ 34    *Answer:*

**43.** 3876 ÷ 68    *Answer:*        **44.** 5913 ÷ 27    *Answer:*

**45.** 3498 ÷ 33    *Answer:*        **46.** 6431 ÷ 59    *Answer:*

**47.** 2835 ÷ 63    *Answer:*        **48.** 7480 ÷ 68    *Answer:*

**E.** Find (a) the quotient and (b) the remainder.

**49.** 89 ÷ 13

*Answer:*    (a) quotient =        (b) remainder =

**50.**  148 ÷ 17

    *Answer:*   (a) quotient =              (b) remainder =

**51.**  536 ÷ 24

    *Answer:*   (a) quotient =              (b) remainder =

**52.**  805 ÷ 42

    *Answer:*   (a) quotient =              (b) remainder =

**53.**  597 ÷ 39

    *Answer:*   (a) quotient =              (b) remainder =

**54.**  3083 ÷ 66

    *Answer:*   (a) quotient =              (b) remainder =

**55.**  8902 ÷ 78

    *Answer:*   (a) quotient =              (b) remainder =

**56.**  9043 ÷ 89

    *Answer:*   (a) quotient =              (b) remainder =

**F.**

**57.**  A worker earns $53 on Monday, $46 on Tuesday, $49 on Wednesday, $38 on Thursday, and $58 on Friday. How much does he earn for this 5-day week? *Answer:*

**58.**  A woman buys six sundresses for $27 each. How much does she spend for these dresses? *Answer:*

**59.**  Nancy pays for a $23 item with a $50 bill. How much change does she receive? *Answer:*

**60.**  A 12-ounce bottle of diet cola contains 84 calories. How many calories are there per ounce? *Answer:*

**61.**  A man drives 150 miles before lunch and an additional 125 miles afterward. Altogether, how many miles does he drive? *Answer:*

**62.**  How many days are there in 9 weeks? *Answer:*

**63.**  An 18-floor building contains 270 apartments. If each floor contains the same number of apartments, how many apartments are there on a floor? *Answer:*

**64.**  Each of the 42 rows in an auditorium contains 36 seats. How many seats are there in this auditorium? *Answer:*

# 3. Averaging

The *average* of several numbers is a notion that has wide application in business, the social sciences, and everyday life. For example, we commonly speak of average income, batting averages, and averages on exams.

## A. CALCULATING AVERAGES

Sometimes we want to consider *several* numbers at a time — perhaps three numbers, or four numbers, or possibly ten numbers. We use the letter $n$ to indicate how many numbers are involved.

DEFINITION

> The **average** of $n$ numbers is the sum of these numbers divided by $n$, that is,
>
> $$\frac{\text{the sum}}{n}$$

**Example 1▶**   Find the average of 66, 78, and 96.

*Solution.*  Here $n = 3$. First add these 3 numbers.

$$\begin{array}{r} 66 \\ 78 \\ \underline{96} \\ 240 \end{array}$$

Now divide by 3.

$$\frac{240}{3} = 80$$

that is,

$$\frac{66 + 78 + 96}{3} = \frac{240}{3} = 80$$

The average of the 3 numbers is 80.                                    ◄

**Example 2** ▶   Find the average of 40, 50, 52, and 56.

*Solution.* Here $n = 4$. First add these four numbers.

$$\begin{array}{r} 40 \\ 50 \\ 52 \\ \underline{56} \\ 198 \end{array}$$

Now divide by 4.

$$\frac{198}{4} = \frac{198.0}{4} = 49.5$$

In other words,

$$\frac{40 + 50 + 52 + 56}{4} = \frac{198}{4} = 49.5$$

Thus the average of the 4 numbers is 49.5.                             ◄

*Try the exercises for Topic A on page 25.*

## B. WORD PROBLEMS

**Example 3** ▶   A student had scores of 86, 68, 70, and 76 on her hour exams. Find the average of these grades.

*Solution.* There were four exams. Thus $n = 4$. Add the 4 grades.

$$\begin{array}{r} 86 \\ 68 \\ 70 \\ \underline{76} \\ 300 \end{array}$$

Now divide by 4.

$$\frac{300}{4} = 75$$

The average of these four grades is 75.                                ◄

**Example 4 ▶**  Suppose that in Example 3, the student receives a 0 on her fifth hour exam. Find the average of her grades of 86, 68, 70, 76, and 0.

*Solution.*  There were 5 exams. (The 0 counts as one of the exam scores.) Thus $n = 5$. The sum is the same as before, namely, 300. But now divide by 5, instead of by 4.

$$\frac{300}{5} = 60$$

The average of these five grades is 60.                                  ◀

*Try the exercises for Topic B on page 26.*

## C. MULTIPLE CHOICE

**Example 5 ▶**  The average of 43, 49, 57, and 71 is

    (A) 53             (B) 54             (C) 55

    (D) 44             (E) 220

*Solution.*  Here $n = 4$.

$$\begin{array}{r} 43 \\ 49 \\ 57 \\ \underline{71} \\ 220 \end{array} \qquad \frac{220}{4} = 55$$

The average of these numbers is 55. Thus (C) is the correct choice.          ◀

    Some averaging problems use algebraic methods. These problems will be discussed in Section 34.

*Try the exercises for Topic C on page 26.*

## EXERCISES

**A.** *Find the average of the given numbers.*

| | | | | |
|---|---|---|---|---|
| 1. | 14, 16, 21 | *Answer:* | 2. | 71, 79, 81   *Answer:* |
| 3. | 37, 62, 84 | *Answer:* | 4. | 92, 105, 112   *Answer:* |
| 5. | 29, 36, 41, 54 | *Answer:* | 6. | 37, 41, 44, 46   *Answer:* |
| 7. | 60, 68, 73, 87 | *Answer:* | 8. | 20, 34, 48, 62   *Answer:* |
| 9. | 77, 93, 98, 100 | *Answer:* | 10. | 59, 83, 89, 93   *Answer:* |
| 11. | 66, 72, 0, 80 | *Answer:* | 12. | 53, 59, 69, 79   *Answer:* |
| 13. | 58, 82, 83, 91 | *Answer:* | 14. | 94, 97, 99, 101   *Answer:* |
| 15. | 37, 42, 44, 51, 61 | *Answer:* | 16. | 28, 40, 51, 52, 59   *Answer:* |

**B.**

17. Melissa had scores of 82, 84, and 92 on her Spanish exams. Find the average of these grades.    *Answer:*

18. Pablo had grades of 76, 81, 57, and 93 on his math quizzes. Find the average of these grades.    *Answer:*

19. In their first four games the Knicks scored 92 points, 94 points, 102 points, and 96 points. Find the average number of points they scored.    *Answer:*

20. The heights of four first-graders are 46 inches, 49 inches, 50 inches, and 51 inches. Find the average height of these children.    *Answer:*

21. A saleswoman makes 12 sales on Monday, 13 sales on Tuesday, 8 sales on Wednesday, and 15 sales on Thursday. Find the average number of sales she makes per day for these four days.    *Answer:*

22. On four successive nights a salesman spends $38, $44, $39, and $47 per night for his motel room. What is the average cost per night for his room?
*Answer:*

23. The average monthly rainfall for a town is as follows: January: 17 centimeters; February: 19 centimeters; March: 25 centimeters; April: 27 centimeters. Find the average monthly rainfall for the town during this period.
*Answer:*

24. In four practice tests, Ann answered 46 questions correctly, then 54 correctly, then 60 correctly, then 64 correctly. What was the average number of questions she answered correctly?    *Answer:*

**C.** *Draw a circle around the correct letter.*

**25.** The average of 17, 23, and 26 is

   (A) 22          (B) 23          (C) 24          (D) 20.6          (E) 68

**26.** The average of 59, 93, and 97 is

   (A) 249          (B) 24.9          (C) 90          (D) 89          (E) 83

**27.** The average of 20, 50, and 59 is

   (A) 49          (B) 129          (C) 43          (D) 34          (E) 34.25

**28.** The average of 59, 61, and 72 is

   (A) 60          (B) 65          (C) 61          (D) 64          (E) 192

**29.** The average of 40, 42, 45, and 49 is

   (A) 42          (B) 43          (C) 44          (D) 45          (E) 176

30. The average of 39, 42, 48, and 51 is

    (A) 42     (B) 45     (C) 48     (D) 50     (E) 180

31. The average of 57, 72, 79, and 80 is

    (A) 288     (B) 28.8     (C) 70     (D) 72     (E) 96

32. The average of 45, 65, 70, and 90 is

    (A) 70     (B) 80     (C) 67.5     (D) 78.5     (E) 77.5

33. A student has scores of 80, 82, 60, and 94 on her biology exams. What was her average score?

    (A) 80     (B) 79     (C) 81     (D) 82     (E) 4

34. A student scored 66, 70, 0, and 84 on his math tests. What was his average score?

    (A) 70     (B) 220     (C) 73.3     (D) 55     (E) 65

35. On different trips the time it takes a train to travel between two stations is 48 minutes, 46 minutes, 47 minutes, and 54 minutes. What is the average number of minutes per trip?

    (A) 38.75     (B) 45     (C) 48.75     (D) 47.7     (E) 50

36. The cost of producing four items is $52, $55, $56, and $61. The average cost per item is

    (A) $55     (B) $55.50     (C) $56     (D) $60     (E) $224

# 4.  Adding and Subtracting Units

It is important to know how to add or subtract units, such as hours and minutes or pounds and ounces.

**A. TIME**

```
1 hour = 60 minutes
1 minute = 60 seconds
```

To add or subtract time units, you may have to change seconds to minutes or minutes to hours.

**Example 1 ▶**    4 hours  33 minutes
                  + 6 hours  29 minutes

*Solution.*  Begin by adding the smaller units.

```
    33 minutes
  + 29 minutes
    62 minutes
```

Because 62 minutes is more than 1 hour, change to hours and minutes:

$$62 \text{ minutes} = 60 \text{ minutes} + 2 \text{ minutes}$$

$$= 1 \text{ hour}  2 \text{ minutes}$$

Thus

```
    4 hours  33 minutes
  + 6 hours  29 minutes
   10 hours  62 minutes
  + 1 hour    2 minutes
   11 hours   2 minutes
```

**Example 2 ▶**    8 minutes  30 seconds
                  - 5 minutes  40 seconds

*Solution.*  Begin at the right with the smaller units. Because you are subtracting 40 seconds from 30 seconds, borrow 1 minute, or 60 seconds, so that instead of 8 minutes 30 seconds, you have 7 minutes 90 seconds.

```
  7 minutes    90 seconds
  8 minutes    30 seconds
- 5 minutes    40 seconds
  2 minutes    50 seconds
```

Alternatively, you could write

```
                90 seconds
  8 minutes     30 seconds
  6 minutes
- 5 minutes     40 seconds
  2 minutes     50 seconds
```

**Example 3 ▶**  A baseball game started at 2:30 P.M. and ended at 5:15 P.M. How long did it last?

*Solution.* Write

| 2:30 P.M. | as | 2 hours  30 minutes |
|-----------|-----|---------------------|
| 5:15 P.M. | as | 5 hours  15 minutes |

Subtract 2 hours 30 minutes from 5 hours 15 minutes.

$$
\begin{array}{rr}
4 \text{ hours} & 75 \text{ minutes} \\
\text{5 hours} & \text{15 minutes} \\
- 2 \text{ hours} & 30 \text{ minutes} \\
\hline
2 \text{ hours} & 45 \text{ minutes}
\end{array}
$$

The game lasted 2 hours and 45 minutes.    ◀

*Try the exercises for Topic A on page 30.*

## B. WEIGHT

> 1 pound = 16 ounces

**Example 4 ▶**  Add 9 pounds 12 ounces and 6 pounds 8 ounces.

*Solution.*

$$
\begin{array}{rr}
9 \text{ pounds} & 12 \text{ ounces} \\
+ 6 \text{ pounds} & 8 \text{ ounces} \\
\hline
15 \text{ pounds} & \text{20 ounces} \\
+ 1 \text{ pound} & 4 \text{ ounces} \\
\hline
16 \text{ pounds} & 4 \text{ ounces}
\end{array}
$$

Note that here you must convert 20 ounces to 1 pound 4 ounces.    ◀

In the *metric system*, the basic unit of weight is called a **gram**. There are about 454 grams to a pound. A paper clip weighs about 1 gram.

> 1 kilogram = 1000 grams
> 1 gram = 100 centigrams

**Example 5 ▶**  Subtract 3 kilograms 650 grams from 5 kilograms 500 grams.

*Solution.*    Borrow 1 kilogram, or 1000 grams.

$$
\begin{array}{rr}
4 \text{ kilograms} & 1500 \text{ grams} \\
\text{5 kilograms} & \text{500 grams} \\
- 3 \text{ kilograms} & 650 \text{ grams} \\
\hline
1 \text{ kilogram} & 850 \text{ grams}
\end{array}
$$
◀

*Try the exercises for Topic B on page 31.*

## C. LENGTH

```
┌─────────────────────┐
│  1 foot = 12 inches │
└─────────────────────┘
```

**Example 6** ▶        4 feet    9 inches
                    − 1 foot  11 inches

*Solution.* Borrow 1 foot, or 12 inches.

          3 feet   21 inches
          ~~4 feet~~   ~~9 inches~~
        − 1 foot  11 inches
          2 feet   10 inches                                    ◄

In the metric system, the basic unit of length is called a **meter**. One meter is approximately 39.37 inches.

```
┌──────────────────────────────────┐
│  1 kilometer = 1000 meters       │
│  1 meter = 100 centimeters       │
└──────────────────────────────────┘
```

**Example 7** ▶        8 meters   45 centimeters
                    +  6 meters   95 centimeters
                      14 meters   ~~140 centimeters~~
                    +  1 meter    40 centimeters
                      15 meters   40 centimeters                ◄

*Try the exercises for Topic C on page 32.*

## EXERCISES

**A.** *Add or subtract.*

1.      5 hours 24 minutes          2.      6 hours 12 minutes
     +  3 hours 28 minutes               +  3 hours 48 minutes


3.      4 hours 20 minutes          4.      9 minutes 15 seconds
     +  3 hours 50 minutes               +  5 minutes 55 seconds


5.      6 hours  4 minutes          6.      8 minutes 21 seconds
     +  5 hours 57 minutes               +  7 minutes 49 seconds


7.      8 hours 25 minutes          8.      3 hours 20 minutes
     −  4 hours  5 minutes               −  1 hour  20 minutes


9.     10 hours  5 minutes         10.      6 hours 15 minutes
     −  4 hours 35 minutes               −  4 hours 30 minutes


11.     7 hours 40 minutes         12.     15 hours 13 minutes
     −  2 hours 48 minutes               −  6 hours 40 minutes

13.    10 hours 15 minutes
    - 6 hours 36 minutes

14.    9 minutes 40 seconds
    - 5 minutes 55 seconds

15.    13 minutes 28 seconds
    - 8 minutes 37 seconds

16.    11 hours 5 minutes
    - 6 hours 6 minutes

17. A show started at 8:40 P.M. and ended at 11:05 P.M. How long did it last? *Answer:*

18. Diane leaves her mother's house at 3:35 P.M. and arrives home at 5:20 P.M. How long does she travel?  *Answer:*

19. A movie started at 7:40 P.M. and ended at 9:25 P.M. How long did it last? *Answer:*

20. A flight left at 8:42 A.M. and arrived at 11:35 A.M. How long did it last? *Answer:*

21. A lecture began at 8:05 P.M. and ended at 9:50 P.M. How long did it last? *Answer:*

22. A meeting began at 11:10 A.M. and ended at 1:05 P.M. How long did it last? *Answer:*

**B.** *Add.*

23.    6 pounds 10 ounces
   + 5 pounds  6 ounces

24.    9 pounds 8 ounces
   + 3 pounds 9 ounces

25.    12 pounds 10 ounces
   +  9 pounds 12 ounces

26.    7 pounds 13 ounces
   + 8 pounds 11 ounces

27.    10 kilograms 120 grams
   +  7 kilograms 380 grams

28.    5 kilograms 600 grams
   + 4 kilograms 700 grams

29.    8 grams 75 centigrams
   + 4 grams 52 centigrams

30.    12 grams 96 centigrams
   + 14 grams 89 centigrams

*Subtract.*

31.    5 pounds 12 ounces
   - 2 pounds 10 ounces

32.    9 pounds 6 ounces
   - 3 pounds 9 ounces

33.    7 pounds
   - 4 pounds 12 ounces

34.    15 pounds  5 ounces
   -  5 pounds 15 ounces

35.    12 pounds 12 ounces
   -  8 pounds 14 ounces

36.    9 pounds  7 ounces
   - 6 pounds 15 ounces

37.    12 kilograms 390 grams
   -  7 kilograms 165 grams

38.    8 kilograms  45 grams
   - 4 kilograms 132 grams

39.     10 grams 17 centigrams
     −  6 grams 62 centigrams

40.     9 grams 12 centigrams
     −      95 centigrams

41.  Add 5 pounds 6 ounces and 9 pounds 11 ounces.     *Answer:*

42.  Add 17 pounds 12 ounces and 8 pounds 13 ounces.     *Answer:*

43.  Subtract 5 pounds 10 ounces from 10 pounds 5 ounces.     *Answer:*

44.  Subtract 11 pounds 12 ounces from 16 pounds.     *Answer:*

45.  Subtract 3 kilograms 500 grams from 10 kilograms.     *Answer:*

46.  Subtract 7 grams 12 centigrams from 9 grams.     *Answer:*

**C.** *Add.*

47.     5 feet 6 inches
     + 3 feet 6 inches

48.     9 feet  2 inches
     + 3 feet 11 inches

49.     7 feet 8 inches
     + 3 feet 9 inches

50.     6 feet  5 inches
     + 3 feet 11 inches

51.     6 kilometers 450 meters
     + 3 kilometers 550 meters

52.     8 meters 25 centimeters
     + 6 meters 85 centimeters

*Subtract.*

53.     7 feet 4 inches
     − 3 feet 5 inches

54.     8 feet 2 inches
     − 2 feet 8 inches

55.     9 feet
     − 4 feet 9 inches

56.     7 feet  3 inches
     − 2 feet 11 inches

57.     5 kilometers 15 meters
     − 2 kilometers 60 meters

58.     12 kilometers 10 meters
     − 10 kilometers 12 meters

59.     3 meters 47 centimeters
     − 2 meters 58 centimeters

60.     15 meters 45 centimeters
     −  5 meters 85 centimeters

# 5.  Adding and Subtracting Fractions

In this section the notion of a fraction is introduced. You will learn how to add and subtract fractions with the same denominator and with different denominators.

## A.  LOWEST TERMS

DEFINITION

**Fractions** are numbers, such as $\frac{1}{5}$ and $\frac{3}{10}$, that are written as quotients. For the fraction $\frac{1}{5}$, 1 is its **numerator** and 5 is its **denominator**. A fraction, such as $\frac{2}{3}$, in which the numerator is smaller than the denominator, is called a **proper fraction**. And a fraction, such as $\frac{3}{2}$ or $\frac{4}{4}$, in which the numerator is larger than or the same as the denominator, is called an **improper fraction**.

Every integer can be written as a fraction with denominator 1. For example, $4 = \frac{4}{1}$.

The fraction $\frac{2}{4}$ is not in its simplest form. It can be **reduced to lowest terms** by dividing the numerator and denominator by 2. Thus

$$\frac{2}{4} = \frac{\overset{1}{\cancel{2} \times 1}}{\underset{1}{\cancel{2} \times 2}} = \frac{1}{2}$$

Note that $\frac{2}{4}$ can be obtained from $\frac{1}{2}$ by multiplying the numerator and denominator of $\frac{1}{2}$ by 2. The fractions $\frac{2}{4}$ and $\frac{1}{2}$ are called **equivalent fractions**.

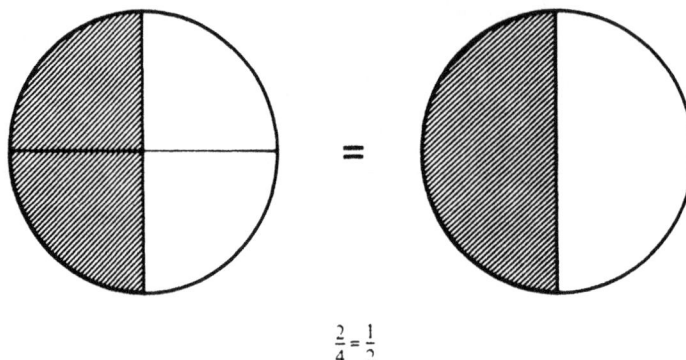

$$\tfrac{2}{4} = \tfrac{1}{2}$$

To reduce a fraction to lowest terms, divide by all factors common to both the numerator and denominator.

**Example 1** ▶    Reduce each fraction to lowest terms.

(a) $\frac{12}{18}$        (b) $\frac{25}{40}$

*Solution.*  Look for factors common to both the numerator and denominator of each fraction.

(a) $\frac{12}{18} = \frac{\overset{1}{\cancel{6} \times 2}}{\underset{1}{\cancel{6} \times 3}} = \frac{2}{3}$        (b) $\frac{25}{40} = \frac{\overset{1}{\cancel{5} \times 5}}{\underset{1}{\cancel{5} \times 8}} = \frac{5}{8}$    ◀

*Try the exercises for Topic A on page 39.*

## B.  ADDING AND SUBTRACTING FRACTIONS WITH THE SAME DENOMINATOR

> *To add or subtract fractions with the <u>same</u> denominator D:*
>
> 1.  Add or subtract the numerators.
> 2.  The denominator is $D$.
> 3.  Reduce to lowest terms, if necessary.

This procedure is illustrated in the accompanying figure.

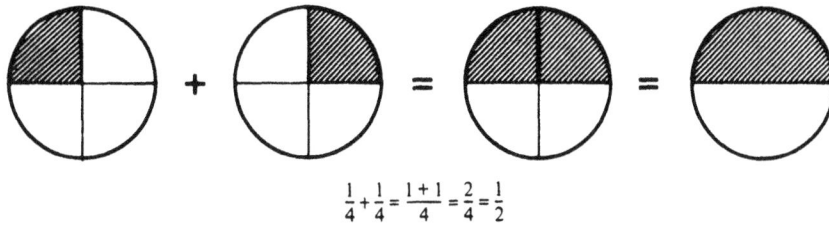

$$\tfrac{1}{4}+\tfrac{1}{4}=\tfrac{1+1}{4}=\tfrac{2}{4}=\tfrac{1}{2}$$

**Example 2 ▶** Add.    $\dfrac{1}{5}+\dfrac{2}{5}$

*Solution.* Both fractions have the same denominator, 5.

Add the numerators.
$$\dfrac{1}{5}+\dfrac{2}{5}=\dfrac{\overbrace{1+2}}{5}=\dfrac{3}{5}$$
Write the denominator 5.

Here $\dfrac{3}{5}$ is already in lowest terms because there are no (positive) factors (other than 1) common to the numerator and denominator.    ◀

**Example 3 ▶** Subtract.    $\dfrac{3}{8}-\dfrac{1}{8}$

*Solution.*

$$\dfrac{3}{8}-\dfrac{1}{8}=\dfrac{3-1}{8}=\dfrac{2}{8}=\dfrac{1}{4}$$

Here  $\dfrac{2}{8}=\dfrac{\overset{1}{\cancel{2}}\times 1}{\underset{1}{\cancel{2}}\times 4}=\dfrac{1}{4}$    ◀

*Try the exercises for Topic B on page 39.*

## C. LEAST COMMON DENOMINATORS

In order to add fractions with *different* denominators, you must first express each of them with the same denominator. To this end, we consider some preliminary notions.

Recall that 6 is a *multiple* of both 2 and 3 because $2\times 3=6$.

**DEFINITION**

> The **least common multiple** of two (or more) integers is the *smallest* positive integer that is a multiple of each of them.

Clearly, none of the integers 1, 2, 3, 4, or 5 is a multiple of *both 2 and 3*. Thus 6 is the least common multiple of 2 and 3.

> *To find the least common multiple of two integers*, first
> write the *smallest* (positive) multiples of each of them. Then
> find the smallest number that is on both lists. This is the
> *least common multiple.*[1]

For example, to find the least common multiple of 4 and 6, observe that
the multiples of 4 are

$$4, \ 8, \ \boxed{12}, \ 16, \ 20, \ 24, \ \text{and so on}$$

and the multiples of 6 are

$$6, \ \boxed{12}, \ 18, \ 24, \ \text{and so on}$$

The smallest number that is on both lists is 12; thus 12 is the least common
multiple of 4 and 6.

**Example 4** ▶  Find the least common multiple of 12 and 16.

*Solution.*  List the multiples of 12 and 16.

$$12, \ 24, \ 36, \ \boxed{48}, \ 60, \text{and so on}$$

$$16, \ 32, \ \boxed{48}, \ 64, \text{and so on}$$

The smallest number that is on both lists is 48. Thus the least common
multiple of 12 and 16 is 48.                                        ◀

DEFINITION

> The **least common denominator** (*lcd*) of two (or more)
> fractions is the least common *multiple* of the individual
> denominators.

For example, the least common denominator of $\frac{2}{3}$ and $\frac{1}{4}$ is the least

common multiple of 3 and 4, or 12. To add or subtract $\frac{2}{3}$ and $\frac{1}{4}$, first

write them as *equivalent* fractions with 12, the *lcd*, as denominator.  To

express $\frac{2}{3}$ with denominator 12, note that $12 = 4 \times 3$. Thus multiply the

numerator and denominator of $\frac{2}{3}$ by 4.

$$\frac{2}{3} = \frac{4 \times 2}{4 \times 3} = \frac{8}{12}$$

To express $\frac{1}{4}$ with denominator 12, note that $12 = 3 \times 4$, and multiply the

numerator and denominator of $\frac{1}{4}$ by 3.

[1] This method is only useful for comparatively simple numbers.

$$\frac{1}{4} = \frac{3 \times 1}{3 \times 4} = \frac{3}{12}$$

**Example 5 ▶**   (a) Find the *lcd* of $\frac{5}{12}$ and $\frac{2}{15}$.

(b) Write equivalent fractions with this *lcd* as denominator.

***Solution.***

(a) First find the least common multiple of the individual denominators, 12 and 15.

12, 24, 36, 48, $\boxed{60}$, 72, and so on

15, 30, 45, $\boxed{60}$, 75, and so on

The smallest number that is on both lists is 60. This is the least common multiple of 12 and 15, and therefore 60 is the *lcd* of $\frac{5}{12}$ and $\frac{2}{15}$.

(b) To write $\frac{5}{12}$ with denominator 60, multiply the numerator and denominator by 5.

$$\frac{5}{12} = \frac{5 \times 5}{5 \times 12} = \frac{25}{60}$$

(c) To write $\frac{2}{15}$ with denominator 60, multiply the numerator and denominator by 4.

$$\frac{2}{15} = \frac{4 \times 2}{4 \times 15} = \frac{8}{60}$$   ◀

*Try the exercises for Topic C on page 40.*

## D.  ADDING AND SUBTRACTING FRACTIONS WITH DIFFERENT DENOMINATORS

> *To add or subtract fractions with <u>different denominators</u>:*
>
> 1. Find their *lcd*.
> 2. Then find equivalent fractions with this *lcd* as denominator.
> 3. Add or subtract the numerators.
> 4. Reduce the resulting fraction to lowest terms, if necessary.

This procedure is illustrated in the accompanying figure.

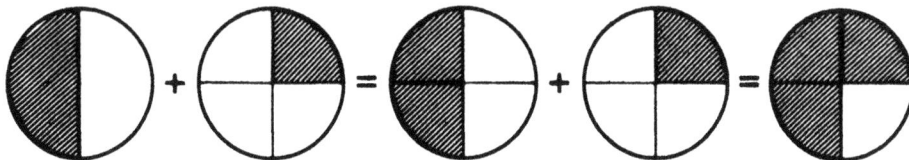

$$\frac{1}{2} + \frac{1}{4} = \frac{2}{4} + \frac{1}{4} = \frac{3}{4}$$

**Example 6 ▶**   Add.    $\dfrac{5}{6} + \dfrac{3}{5}$

**Solution.**  The *lcd* of $\dfrac{5}{6}$ and $\dfrac{3}{5}$ is 30, the least common multiple of 6 and 5.

$$\frac{5}{6} = \frac{5 \times 5}{5 \times 6} = \frac{25}{30}$$

$$\frac{3}{5} = \frac{6 \times 3}{6 \times 5} = \frac{18}{30}$$

$$\frac{5}{6} + \frac{3}{5} = \frac{25}{30} + \frac{18}{30} = \frac{25 + 18}{30} = \frac{43}{30}$$

The numerator and denominator of $\dfrac{43}{30}$ have no factor in common. Thus

$$\frac{5}{6} + \frac{3}{5} = \frac{43}{30} \qquad ◀$$

**Example 7 ▶**   Subtract.    $\dfrac{11}{20} - \dfrac{5}{12}$

**Solution.**  To find the least common multiple of 20 and 12, consider the smallest positive multiple of each.

20,  40,  $\boxed{60}$,  80,  100, and so on

12,  24,  36,  48,  $\boxed{60}$, and so on

Thus 60 is the least common multiple of 20 and 12, and is thus the *lcd* of $\dfrac{11}{20}$ and $\dfrac{5}{12}$.

$$\frac{11}{20} = \frac{3 \times 11}{3 \times 20} = \frac{33}{60}$$

$$\frac{5}{12} = \frac{5 \times 5}{5 \times 12} = \frac{25}{60}$$

$$\frac{11}{20} - \frac{5}{12} = \frac{33}{60} - \frac{25}{60}$$

$$= \frac{33 - 25}{60}$$

$$= \frac{8}{60} \qquad \text{(Divide numerator and denominator by 4.)}$$

$$= \frac{2}{15} \qquad ◀$$

*Try the exercises for Topic D on page 40.*

## E. MULTIPLE CHOICE

**Example 8** ▶  $\frac{1}{9} + \frac{3}{5} =$

(A) $\frac{4}{14}$    (B) $\frac{2}{7}$    (C) $\frac{4}{45}$    (D) $\frac{32}{45}$    (E) $\frac{4}{9}$

*Solution.* The *lcd* of $\frac{1}{9}$ and $\frac{3}{5}$ is 45.

$$\frac{1}{9} = \frac{5 \times 1}{5 \times 9} = \frac{5}{45}$$

$$\frac{3}{5} = \frac{9 \times 3}{9 \times 5} = \frac{27}{45}$$

$$\frac{1}{9} + \frac{3}{5} = \frac{5}{45} + \frac{27}{45}$$

$$= \frac{5 + 27}{45}$$

$$= \frac{32}{45}$$

This is in lowest terms. The correct choice is (D).    ◄

*Try the exercises for Topic E on page 41.*

## EXERCISES

**A.** *Reduce each fraction to lowest terms.*

1. $\frac{7}{14}$    *Answer:*    2. $\frac{4}{20}$    *Answer:*

3. $\frac{15}{27}$    *Answer:*    4. $\frac{35}{50}$    *Answer:*

5. $\frac{36}{54}$    *Answer:*    6. $\frac{84}{98}$    *Answer:*

7. $\frac{108}{144}$    *Answer:*    8. $\frac{128}{192}$    *Answer:*

**B.** *Add or subtract.*

9. $\frac{2}{7} + \frac{3}{7}$    *Answer:*    10. $\frac{5}{9} + \frac{1}{9}$    *Answer:*

11. $\frac{4}{11} + \frac{7}{11}$    *Answer:*    12. $\frac{5}{8} - \frac{3}{8}$    *Answer:*

13. $\frac{7}{6} - \frac{5}{6}$    *Answer:*    14. $\frac{9}{10} + \frac{3}{10}$    *Answer:*

15. $\frac{17}{20} - \frac{3}{20}$    *Answer:*        16. $\frac{31}{42} - \frac{11}{42}$    *Answer:*

17. $\frac{25}{64} + \frac{7}{64}$    *Answer:*        18. $\frac{53}{98} - \frac{5}{98}$    *Answer:*

**C.** *Find the least common multiple of the given integers.*

19. 5 and 7    *Answer:*        20. 4 and 8    *Answer:*

21. 6 and 8    *Answer:*        22. 10 and 15    *Answer:*

23. 18 and 30    *Answer:*        24. 14 and 49    *Answer:*

25. 36 and 64    *Answer:*        26. 48 and 72    *Answer:*

*In Exercises 27–36:*

(a) *Find the lcd of the given fractions.*

(b) *Write equivalent fractions with this lcd as denominator.*

27. $\frac{1}{2}$ and $\frac{1}{4}$    *Answer:*    (a)        (b)

28. $\frac{2}{3}$ and $\frac{1}{5}$    *Answer:*    (a)        (b)

29. $\frac{3}{4}$ and $\frac{5}{6}$    *Answer:*    (a)        (b)

30. $\frac{3}{10}$ and $\frac{4}{15}$    *Answer:*    (a)        (b)

31. $\frac{2}{9}$ and $\frac{1}{12}$    *Answer:*    (a)        (b)

32. $\frac{4}{5}$ and $\frac{9}{20}$    *Answer:*    (a)        (b)

33. $\frac{7}{12}$ and $\frac{3}{16}$    *Answer:*    (a)        (b)

34. $\frac{1}{18}$ and $\frac{5}{24}$    *Answer:*    (a)        (b)

35. $\frac{1}{30}$ and $\frac{7}{40}$    *Answer:*    (a)        (b)

36. $\frac{1}{48}$ and $\frac{5}{36}$    *Answer:*    (a)        (b)

**D.** *Add or subtract.*

37. $\frac{1}{2} + \frac{1}{3} =$        38. $\frac{1}{4} + \frac{1}{8} =$        39. $\frac{3}{5} - \frac{1}{3} =$

40. $\frac{7}{10} - \frac{1}{2} =$    41. $\frac{2}{7} + \frac{1}{9} =$    42. $\frac{3}{8} - \frac{1}{6} =$

43. $\frac{5}{8} + \frac{3}{10} =$    44. $\frac{7}{9} - \frac{1}{6} =$    45. $\frac{7}{12} - \frac{1}{10} =$

46. $\frac{3}{8} + \frac{2}{3} =$    47. $\frac{11}{12} - \frac{5}{8} =$    48. $\frac{7}{10} + \frac{3}{20} =$

49. $\frac{9}{25} - \frac{3}{10} =$    50. $\frac{1}{36} + \frac{7}{24} =$    51. $\frac{5}{18} + \frac{5}{12} =$

52. $\frac{9}{40} - \frac{2}{25} =$    53. $\frac{1}{4} + \frac{2}{3} =$    54. $\frac{3}{4} + \frac{1}{8} =$

55. $\frac{1}{10} - \frac{1}{20} =$

E.  *Draw a circle around the correct letter.*

56. $\frac{1}{4} - \frac{1}{9} =$

   (A) $\frac{1}{5}$    (B) $\frac{1}{36}$    (C) $\frac{5}{36}$    (D) $\frac{1}{6}$    (E) $\frac{13}{36}$

57. $\frac{3}{4} + \frac{1}{12} =$

   (A) $\frac{4}{16}$    (B) $\frac{4}{48}$    (C) $\frac{5}{6}$    (D) $\frac{11}{12}$    (E) 1

58. $\frac{7}{10} - \frac{1}{5} =$

   (A) $\frac{1}{2}$    (B) $\frac{6}{5}$    (C) $\frac{3}{5}$    (D) $\frac{7}{5}$    (E) $\frac{1}{5}$

59. $\frac{4}{5} + \frac{1}{8} =$

   (A) $\frac{37}{40}$    (B) $\frac{5}{13}$    (C) $\frac{4}{13}$    (D) $\frac{5}{40}$    (E) $\frac{33}{40}$

60. $\frac{4}{7} - \frac{1}{2} =$

   (A) $\frac{3}{5}$    (B) $\frac{3}{7}$    (C) $\frac{3}{14}$    (D) $\frac{1}{14}$    (E) $\frac{1}{2}$

61. $\frac{5}{12} + \frac{3}{8} =$

(A) $\frac{2}{5}$    (B) $\frac{1}{3}$    (C) $\frac{15}{96}$    (D) $\frac{1}{12}$    (E) $\frac{19}{24}$

62.  $\frac{7}{9} - \frac{2}{3} =$

(A) $\frac{1}{9}$    (B) $\frac{5}{6}$    (C) $\frac{5}{9}$    (D) $\frac{14}{27}$    (E) $\frac{1}{3}$

63.  $\frac{5}{7} - \frac{1}{6} =$

(A) $\frac{23}{42}$    (B) $4$    (C) $\frac{29}{42}$    (D) $\frac{2}{21}$    (E) $\frac{4}{21}$

64.  $\frac{3}{10} + \frac{1}{12} =$

(A) $\frac{2}{11}$    (B) $\frac{3}{22}$    (C) $\frac{1}{30}$    (D) $\frac{23}{60}$    (E) $\frac{19}{60}$

65.  $\frac{5}{6} + \frac{3}{8} =$

(A) $\frac{4}{7}$    (B) $\frac{1}{6}$    (C) $\frac{29}{24}$    (D) $\frac{1}{3}$    (E) $\frac{23}{24}$

66.  $\frac{7}{10} + \frac{4}{15} =$

(A) $\frac{11}{25}$    (B) $\frac{29}{30}$    (C) $\frac{5}{6}$    (D) $\frac{1}{2}$    (E) $\frac{11}{15}$

67.  Ed eats $\frac{1}{4}$ of an apple pie and Joe eats $\frac{3}{8}$ of the pie. How much of the pie remains?

(A) $\frac{1}{2}$    (B) $\frac{5}{8}$    (C) $\frac{3}{8}$    (D) $\frac{1}{8}$    (E) $\frac{1}{4}$

68.  A stock goes up $\frac{3}{4}$ of a point one day and it falls $\frac{5}{8}$ of a point the next day. What is its gain or loss for the two days?

(A) It gains $\frac{1}{8}$ of a point.        (B) It loses $\frac{1}{8}$ of a point.

(C) It loses $\frac{1}{4}$ of a point.        (D) It gains one point.

(E) It gains $\frac{3}{8}$ of a point.

69.  A 1-meter pole is immersed in a pond so that $\frac{1}{5}$ of the pole is in bottom sand and $\frac{1}{10}$ is in water. What is the length of the pole above water?

(A)  15 centimeters          (B)  30 centimeters

(C)  40 centimeters          (D)  60 centimeters

(E)  70 centimeters

70.  On a $\frac{3}{4}$ mile trip, Judy bikes for $\frac{2}{3}$ of a mile. How much of the trip remains?

*Answer:*

# 6.  Multiplying and Dividing Fractions

It is probably easier to multiply or divide fractions than it is to add or subtract them.

## A. MULTIPLYING FRACTIONS

The accompanying figure illustrates how fractions are multiplied.

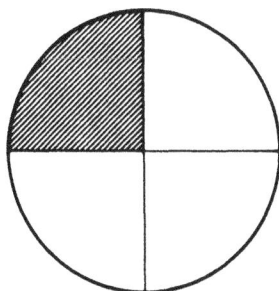

$$\frac{1}{4}$$

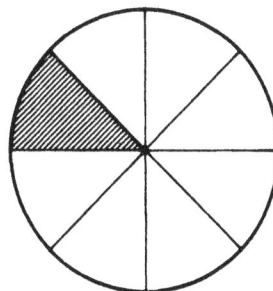

$$\frac{1}{2} \times \frac{1}{4} = \frac{1}{8}$$

> *To multiply two fractions:*
> 1. Divide by factors common to the numerators and denominators.
> 2. Multiply the numerators.
> 3. Multiply the denominators.

**Example 1▶**  Multiply.    $\dfrac{3}{5} \times \dfrac{1}{4}$

*Solution.* The numerators and denominators have no factors in common. Thus multiply the numerators and multiply the denominators.

$$\frac{3}{5} \times \frac{1}{4} = \frac{3 \times 1}{5 \times 4} = \frac{3}{20}$$   ◀

When the numerators and denominators have no factors in common, as in Example 1, the rule for multiplying fractions can be expressed as follows:

$$\frac{a}{b} \times \frac{c}{d} = \frac{a \times c}{b \times d}$$

When the numerators and denominators have factors in common, it is best to divide by these common factors before multiplying. You will then be working with smaller numbers.

**Example 2▶**  Multiply.    $\dfrac{3}{10} \times \dfrac{4}{9}$

*Solution.* Divide 3 and 9 by 3. Divide 4 and 10 by 2.

$$\frac{\overset{1}{\cancel{3}}}{\underset{5}{\cancel{10}}} \times \frac{\overset{2}{\cancel{4}}}{\underset{3}{\cancel{9}}} = \frac{1 \times 2}{5 \times 3} = \frac{2}{15}$$   ◀

*Try the exercises for Topic A on page 45.*

## B. DIVIDING FRACTIONS

To divide fractions,

$$\frac{a}{b} \div \frac{c}{d}$$

invert the divisor $\dfrac{c}{d}$ ) to obtain $\dfrac{d}{c}$, and then multiply. Thus

$$\frac{a}{b} \div \frac{c}{d} = \frac{a}{b} \times \frac{d}{c} = \frac{a \times d}{b \times c}$$

**Example 3** ▶ Divide.    $\frac{1}{5} \div \frac{2}{3}$

**Solution.**

$$\frac{1}{5} \div \frac{2}{3} = \frac{1}{5} \times \frac{3}{2}$$

$$= \frac{1 \times 3}{5 \times 2}$$

$$= \frac{3}{10}$$ ◀

**Example 4** ▶ Divide.    $\frac{9}{16} \div \frac{3}{8}$

**Solution.**

$$\frac{9}{16} \div \frac{3}{8} = \frac{\overset{3}{\cancel{9}}}{\underset{2}{\cancel{16}}} \times \frac{\overset{1}{\cancel{8}}}{\underset{1}{\cancel{3}}} = \frac{3}{2}$$ ◀

*Try the exercises for Topic B on page 46.*

## C. MULTIPLE CHOICE

**Example 5** ▶ $\frac{4}{15} \div \frac{2}{5} =$

(A) $\frac{8}{75}$    (B) $\frac{2}{3}$    (C) $\frac{2}{15}$    (D) $\frac{2}{75}$    (E) $\frac{20}{15}$

**Solution.**

$$\frac{4}{15} \div \frac{2}{5} = \frac{\overset{2}{\cancel{4}}}{\underset{3}{\cancel{15}}} \times \frac{\overset{1}{\cancel{5}}}{\underset{1}{\cancel{2}}} = \frac{2}{3}$$

The correct choice is (B). ◀

*Try the exercises for Topic C on page 46.*

## EXERCISES

**A.** *Multiply.*

1. $\frac{1}{2} \times \frac{1}{3}$    *Answer:*    2. $\frac{3}{4} \times \frac{1}{4}$    *Answer:*

3. $\frac{2}{5} \times \frac{1}{2}$    *Answer:*      4. $\frac{3}{10} \times \frac{1}{3}$    *Answer:*

5. $\frac{5}{7} \times \frac{2}{5}$    *Answer:*      6. $\frac{7}{9} \times \frac{3}{7}$    *Answer:*

7. $\frac{5}{8} \times \frac{4}{15}$    *Answer:*      8. $\frac{7}{10} \times \frac{5}{14}$    *Answer:*

9. $\frac{8}{9} \times \frac{3}{4}$    *Answer:*      10. $\frac{9}{10} \times \frac{5}{3}$    *Answer:*

11. $\frac{3}{7} \times \frac{5}{6}$    *Answer:*      12. $\frac{7}{12} \times \frac{3}{14}$    *Answer:*

13. $\frac{5}{12} \times \frac{1}{15}$    *Answer:*      14. $\frac{7}{20} \times \frac{3}{14}$    *Answer:*

15. $\frac{9}{100} \times \frac{10}{27}$    *Answer:*      16. $\frac{3}{32} \times \frac{12}{13}$    *Answer:*

**B.** *Divide.*

17. $\frac{1}{3} \div \frac{1}{2}$    *Answer:*      18. $\frac{1}{2} \div \frac{1}{3}$    *Answer:*

19. $\frac{1}{4} \div \frac{1}{2}$    *Answer:*      20. $\frac{2}{5} \div \frac{4}{7}$    *Answer:*

21. $\frac{3}{8} \div \frac{9}{10}$    *Answer:*      22. $\frac{5}{6} \div \frac{15}{16}$    *Answer:*

23. $\frac{7}{12} \div \frac{7}{9}$    *Answer:*      24. $\frac{3}{8} \div \frac{9}{2}$    *Answer:*

25. $\frac{4}{9} \div \frac{2}{15}$    *Answer:*      26. $\frac{2}{3} \div \frac{6}{7}$    *Answer:*

27. $\frac{4}{5} \div \frac{8}{25}$    *Answer:*      28. $9 \div \frac{1}{6}$    *Answer:*

29. $\frac{1}{9} \div 6$    *Answer:*      30. $12 \div \frac{2}{3}$    *Answer:*

31. $\frac{5}{144} \div \frac{25}{12}$    *Answer:*      32. $\frac{9}{64} \div \frac{27}{32}$    *Answer:*

**C.** *Draw a circle around the correct letter.*

33. $\frac{2}{7} \times \frac{5}{8} =$

(A) $\frac{7}{15}$      (B) $\frac{5}{28}$      (C) $\frac{1}{4}$      (D) $\frac{25}{78}$      (E) $\frac{16}{35}$

34. $\frac{7}{10} \times \frac{2}{21} =$

    (A) $\frac{7}{100}$     (B) $\frac{9}{210}$     (C) $\frac{1}{15}$     (D) $\frac{147}{20}$     (E) $\frac{9}{31}$

35. $\frac{9}{14} \times \frac{2}{3} =$

    (A) $\frac{3}{7}$     (B) $\frac{7}{3}$     (C) $\frac{27}{28}$     (D) $\frac{9}{42}$     (E) $\frac{11}{42}$

36. $\frac{5}{9} \times \frac{3}{10} =$

    (A) $\frac{1}{6}$     (B) 6     (C) $\frac{50}{27}$     (D) $\frac{27}{50}$     (E) $\frac{8}{90}$

37. $\frac{2}{5} \div \frac{4}{9} =$

    (A) $\frac{8}{45}$     (B) $\frac{9}{10}$     (C) $\frac{10}{9}$     (D) $\frac{45}{8}$     (E) $\frac{38}{45}$

38. $\frac{3}{7} \div \frac{9}{10} =$

    (A) $\frac{10}{21}$     (B) $\frac{21}{10}$     (C) $\frac{27}{70}$     (D) $\frac{10}{7}$     (E) $\frac{7}{30}$

39. $\frac{5}{6} \div \frac{1}{12} =$

    (A) $\frac{5}{72}$     (B) 10     (C) $\frac{1}{10}$     (D) 2     (E) $\frac{11}{12}$

40. $\frac{7}{10} \div \frac{3}{5} =$

    (A) $\frac{1}{10}$     (B) $\frac{7}{6}$     (C) $\frac{6}{7}$     (D) $\frac{21}{50}$     (E) $\frac{50}{21}$

41. $\frac{5}{12} \times \frac{3}{8} =$

    (A) $\frac{5}{32}$     (B) $\frac{11}{8}$     (C) $\frac{10}{9}$     (D) $\frac{9}{10}$     (E) $\frac{19}{24}$

42. $\frac{6}{7} \div 14 =$

    (A) $\frac{3}{49}$     (B) 12     (C) $\frac{1}{12}$     (D) 3     (E) $\frac{2}{49}$

**43.** $\frac{9}{10} \times 15 =$

(A) $\frac{3}{2}$     (B) $\frac{27}{2}$     (C) $\frac{2}{3}$     (D) $\frac{3}{50}$     (E) $\frac{9}{2}$

**44.** $\frac{3}{32} \div \frac{4}{9} =$

(A) $\frac{3}{8}$     (B) $\frac{8}{3}$     (C) $\frac{1}{32}$     (D) $\frac{27}{128}$     (E) $\frac{128}{27}$

**45.** Jerry finds three-fourths of an apple pie in the refrigerator and eats one-sixth of this portion. What fraction of the entire pie does he eat?

(A) $\frac{1}{24}$     (B) $\frac{3}{8}$     (C) $\frac{1}{12}$     (D) $\frac{1}{8}$     (E) $\frac{5}{24}$

**46.** Three-fifths of the students at a college are women. One-tenth of the women are psychology majors. What fraction of the students at this college are women who major in psychology?

(A) $\frac{3}{10}$     (B) $\frac{3}{50}$     (C) $\frac{1}{6}$     (D) $\frac{1}{2}$     (E) $\frac{7}{10}$

# 7.  Mixed Numbers

Mixed numbers involve an integer plus a proper fraction. Here we consider the basic arithmetic operations on mixed numbers.

## A. MIXED NUMBERS AND IMPROPER FRACTIONS

The number

$$4\frac{1}{2} \qquad \text{stands for} \qquad 4 + \frac{1}{2}$$

This is a **mixed number** (an integer plus a proper fraction). Here 4 is called the **integral part** and $\frac{1}{2}$ is called the **fractional part**. Observe that

$$4\frac{1}{2} = 4 + \frac{1}{2}$$
$$= \frac{4}{1} + \frac{1}{2}$$
$$= \frac{8}{2} + \frac{1}{2}$$
$$= \frac{9}{2}$$

Thus the mixed number $4\frac{1}{2}$ can also be expressed as the improper fraction $\frac{9}{2}$.

**Example 1** ▶  Express $4\frac{2}{3}$ as an improper fraction.

*Solution.*

$$4\frac{2}{3} = 4 + \frac{2}{3}$$

$$= \frac{4}{1} + \frac{2}{3}$$

$$= \frac{12}{3} + \frac{2}{3}$$

$$= \frac{14}{3}$$

◀

*Try the exercises for Topic A on page 53.*

## B. ADDING AND SUBTRACTING MIXED NUMBERS

*To add or subtract mixed numbers:*

1. Add or subtract the fractional parts.
2. Add or subtract the integral parts.
3. Simplify, if necessary.

**Example 2 ▶**    Add.    $4\frac{1}{3} + 2\frac{2}{3}$

*Solution.*

$$4 + \frac{1}{3}$$
$$+ 2 + \frac{2}{3}$$
$$\overline{6 + \frac{3}{3}} = 6 + 1 \qquad (\text{because } \frac{3}{3} = 1)$$
$$= 7$$

◀

Sometimes one or both of the mixed numbers must be changed to an equivalent form before adding or subtracting.

**Example 3 ▶**    Subtract.    $5\frac{1}{2} - 2\frac{3}{4}$

*Solution.*  The fractional parts have different denominators. The *lcd* of $\frac{1}{2}$ and $\frac{3}{4}$ is 4. Thus write $\frac{1}{2} = \frac{2}{4}$ and consider

$$5\frac{2}{4}$$
$$- 2\frac{3}{4}$$
$$\overline{\phantom{xxxx}}$$

First subtract the fractional parts. Because $\frac{2}{4}$ is smaller than $\frac{3}{4}$, borrow 1 from the integral part, 5. Thus

$$5\frac{2}{4} = 4 + 1 + \frac{2}{4}$$

$$= 4 + \frac{4}{4} + \frac{2}{4}$$

$$= 4 + \frac{6}{4}$$

Now subtract.

$$4\frac{6}{4}$$
$$- 2\frac{3}{4}$$
$$\overline{\phantom{-} 2\frac{3}{4}}$$

Thus

$$5\frac{1}{2} - 2\frac{3}{4} = 2\frac{3}{4}$$

*Alternate Solution.* Convert to fractions and subtract. Then write the resulting fraction as a mixed number.

$$5\frac{1}{2} - 2\frac{3}{4} = \frac{11}{2} - \frac{11}{4}$$

$$= \frac{22}{4} - \frac{11}{4}$$

$$= \frac{11}{4}$$

$$= \frac{8}{4} + \frac{3}{4} \quad \overset{\text{integral part}}{\underset{\text{fractional part}}{}}$$

$$= 2\frac{3}{4} \qquad \blacktriangleleft$$

*Try the exercises for Topic B on page 53.*

### C. MULTIPLYING AND DIVIDING MIXED NUMBERS

> *To multiply or divide mixed numbers:*
>
> 1. Convert to equivalent fractions.
> 2. Multiply or divide these fractions.
> 3. Write the resulting fraction as a mixed number.

**Example 4 ▶**    Multiply.    $4\frac{1}{5} \times 2\frac{1}{2}$

*Solution.*

$$4\frac{1}{5} = 4 + \frac{1}{5} = \frac{20}{5} + \frac{1}{5} = \frac{21}{5}$$

$$2\frac{1}{2} = 2 + \frac{1}{2} = \frac{4}{2} + \frac{1}{2} = \frac{5}{2}$$

$$4\frac{1}{5} \times 2\frac{1}{2} = \frac{21}{\overset{}{\underset{1}{5}}} \times \frac{\overset{1}{5}}{2}$$

$$= \frac{21}{2}$$

$$= \frac{20}{2} + \frac{1}{2} \quad \text{integral part} \quad \text{fractional part}$$

$$= 10\frac{1}{2} \qquad \qquad \blacktriangleleft$$

**Example 5** ▶  Divide.    $5 \div 3\frac{1}{4}$

*Solution.*  First observe that

$$3\frac{1}{4} = 3 + \frac{1}{4} = \frac{12}{4} + \frac{1}{4} = \frac{13}{4}$$

so that

$$5 \div 3\frac{1}{4} = \frac{5}{1} \div \frac{13}{4}$$

$$= \frac{5}{1} \times \frac{4}{13}$$

$$= \frac{20}{13}$$

$$= \frac{13}{13} + \frac{7}{13}$$

$$= 1\frac{7}{13} \qquad \qquad \blacktriangleleft$$

*Try the exercises for Topic C on page 54.*

## D. MULTIPLE CHOICE

Sometimes the result must be expressed as a fraction, as in Example 6.

**Example 6** ▶  $3\frac{1}{4} - 1\frac{2}{3} =$

(A) $\frac{29}{12}$      (B) $\frac{31}{12}$      (C) $\frac{19}{12}$      (D) 2      (E) $\frac{7}{12}$

*Solution.*  The *lcd* of $\frac{1}{4}$ and $\frac{2}{3}$ is 12.

$$\frac{1}{4} = \frac{3 \times 1}{3 \times 4} = \frac{3}{12}$$

$$\frac{2}{3} = \frac{4 \times 2}{4 \times 3} = \frac{8}{12}$$

Because $\frac{3}{12}$ is smaller than $\frac{8}{12}$, borrow 1 from the integral part, 3.

$$3\frac{1}{4} = 3\frac{3}{12} = 2 + \frac{12}{12} + \frac{3}{12} = 2\frac{15}{12}$$

$$2\frac{15}{12}$$
$$-1\frac{8}{12}$$
$$\overline{1\frac{7}{12}}$$

Now observe that $1\frac{7}{12} = \frac{12}{12} + \frac{7}{12} = \frac{19}{12}$. The correct choice is (C).    ◀

*Try the exercises for Topic D on page 54.*

**EXERCISES**

A.   *Express as an improper fraction.*

1.  $1\frac{1}{2}$        *Answer:*            2.  $2\frac{1}{4}$        *Answer:*

3.  $4\frac{1}{5}$        *Answer:*            4.  $1\frac{3}{4}$        *Answer:*

5.  $5\frac{2}{5}$        *Answer:*            6.  $10\frac{3}{4}$        *Answer:*

7.  $7\frac{2}{3}$        *Answer:*            8.  $9\frac{5}{8}$        *Answer:*

B.   *Add or subtract. Express the result either as an integer, as a mixed number, or as a proper fraction.*

9.  $2\frac{1}{4} + 3\frac{1}{4}$        *Answer:*            10.  $6\frac{2}{3} - 4\frac{1}{3}$        *Answer:*

11.  $4\frac{1}{2} + 1\frac{1}{2}$        *Answer:*            12.  $5\frac{2}{5} - 2\frac{2}{5}$        *Answer:*

13.  $5\frac{1}{2} + 2\frac{1}{4}$        *Answer:*            14.  $6\frac{3}{4} - 1\frac{1}{2}$        *Answer:*

15.  $2\frac{2}{5} + 1\frac{1}{2}$        *Answer:*            16.  $5\frac{2}{3} + 1\frac{1}{6}$        *Answer:*

17. $3\frac{3}{4} + 5\frac{3}{8}$    *Answer:*          18. $10\frac{1}{2} + 7\frac{3}{4}$    *Answer:*

19. $3\frac{1}{2} - 1\frac{3}{4}$    *Answer:*          20. $4\frac{1}{4} - 2\frac{1}{2}$    *Answer:*

21. $7\frac{1}{10} - 4\frac{3}{5}$    *Answer:*          22. $4\frac{2}{9} - 2\frac{2}{3}$    *Answer:*

23. $3\frac{1}{8} - 2\frac{3}{4}$    *Answer:*          24. $5\frac{1}{12} - 4\frac{3}{8}$    *Answer:*

**C.** *Multiply or divide. Express the result either as an integer, as a mixed number, or as a proper fraction.*

25. $4 \times 2\frac{1}{2}$    *Answer:*          26. $8 \times 2\frac{2}{3}$    *Answer:*

27. $2\frac{1}{4} \times 3\frac{1}{2}$    *Answer:*          28. $1\frac{2}{5} \times 1\frac{1}{2}$    *Answer:*

29. $3\frac{3}{4} \times 1\frac{1}{8}$    *Answer:*          30. $2\frac{1}{5} \times 4\frac{1}{4}$    *Answer:*

31. $6 \div 1\frac{1}{2}$    *Answer:*          32. $8 \div 2\frac{3}{4}$    *Answer:*

33. $9\frac{1}{2} \div 3$    *Answer:*          34. $2\frac{1}{4} \div 1\frac{3}{4}$    *Answer:*

35. $5\frac{1}{2} \div 1\frac{5}{8}$    *Answer:*          36. $6\frac{1}{4} \div \frac{3}{8}$    *Answer:*

37. $3\frac{2}{3} \div 1\frac{5}{6}$    *Answer:*          38. $3\frac{3}{5} \div 1\frac{7}{10}$    *Answer:*

*Multiply or divide. Express the result as a fraction.*

39. $2\frac{1}{6} \times 5\frac{1}{4}$    *Answer:*          40. $7\frac{1}{2} \div 3\frac{1}{4}$    *Answer:*

41. $12\frac{1}{2} \div 10\frac{5}{8}$    *Answer:*          42. $7\frac{3}{4} \div 2\frac{1}{12}$    *Answer:*

**D.** *Draw a circle around the correct letter.*

43. $4\frac{1}{2} + 2\frac{3}{4} =$

(A) $6\frac{1}{4}$    (B) $7\frac{1}{4}$    (C) $7\frac{1}{2}$    (D) $6\frac{3}{4}$    (E) $6\frac{7}{8}$

**44.** $3\frac{3}{5} - 2\frac{4}{5} =$

    (A) $\frac{1}{5}$      (B) $1\frac{1}{5}$      (C) $\frac{4}{5}$      (D) $\frac{8}{5}$      (E) $1\frac{2}{5}$

**45.** $10 - 2\frac{1}{7} =$

    (A) $8\frac{1}{7}$      (B) $8$      (C) $7\frac{9}{10}$      (D) $7\frac{6}{7}$      (E) $7$

**46.** $2\frac{1}{4} + 1\frac{1}{6} =$

    (A) $3\frac{1}{10}$      (B) $3\frac{1}{5}$      (C) $3\frac{1}{12}$      (D) $3\frac{5}{12}$      (E) $3\frac{1}{4}$

**47.** $2\frac{1}{4} \times 4\frac{1}{2} =$

    (A) $8\frac{1}{8}$      (B) $6\frac{3}{4}$      (C) $10$      (D) $\frac{81}{8}$      (E) $\frac{1}{2}$

**48.** $3\frac{1}{6} \div 3\frac{1}{3} =$

    (A) $1$      (B) $\frac{1}{2}$      (C) $\frac{1}{6}$      (D) $\frac{1}{18}$      (E) $\frac{19}{20}$

**49.** $8 - 4\frac{3}{4} =$

    (A) $4\frac{1}{4}$      (B) $3\frac{1}{4}$      (C) $\frac{19}{4}$      (D) $\frac{9}{4}$      (E) $\frac{17}{4}$

**50.** $6 \div 2\frac{2}{3} =$

    (A) $16$      (B) $\frac{9}{4}$      (C) $\frac{10}{3}$      (D) $\frac{9}{8}$      (E) $\frac{3}{10}$

**51.** $7\frac{1}{4} - 4\frac{3}{4} =$

    (A) $3\frac{1}{2}$      (B) $3$      (C) $2\frac{1}{2}$      (D) $\frac{11}{4}$      (E) $\frac{5}{4}$

**52.** $2\frac{1}{8} + 5\frac{3}{4} =$

    (A) $8$      (B) $8\frac{1}{8}$      (C) $7\frac{7}{8}$      (D) $8\frac{1}{4}$      (E) $7\frac{3}{32}$

**53.** $5\frac{1}{10} - 2\frac{1}{5} =$

(A) $2\frac{1}{2}$    (B) $3\frac{1}{10}$    (C) $\frac{16}{5}$    (D) $2\frac{9}{10}$    (E) $\frac{19}{10}$

**54.** $2\frac{2}{3} \div 6 =$

(A) $4\frac{1}{3}$    (B) $3\frac{1}{3}$    (C) $12\frac{2}{3}$    (D) 16    (E) $\frac{4}{9}$

**55.** $2\frac{1}{4} \div 1\frac{2}{5} =$

(A) $1\frac{1}{20}$    (B) $\frac{1}{20}$    (C) $\frac{63}{20}$    (D) $\frac{45}{28}$    (E) $1\frac{5}{8}$

**56.** $2\frac{3}{4} - 2\frac{3}{8} =$

(A) 0    (B) $\frac{3}{8}$    (C) $\frac{3}{4}$    (D) $\frac{11}{19}$    (E) $\frac{19}{11}$

**57.** $6\frac{1}{4} - 4\frac{3}{5} =$

(A) 2    (B) $2\frac{2}{5}$    (C) $\frac{27}{20}$    (D) $\frac{33}{20}$    (E) $\frac{23}{5}$

**58.** $3\frac{3}{4} \div 1\frac{1}{12} =$

(A) $3\frac{2}{15}$    (B) $3\frac{2}{13}$    (C) $2\frac{2}{3}$    (D) $\frac{45}{13}$    (E) $\frac{195}{48}$

**59.** A $10\frac{1}{2}$ -yard rope is divided into 4 equal parts. How long is each piece?

(A) $2\frac{1}{2}$ yards    (B) $2\frac{1}{8}$ yards    (C) $2\frac{3}{8}$ yards    (D) $2\frac{5}{8}$ yards

(E) $2\frac{7}{8}$ yards

**60.** A stock which sold at $64\frac{1}{8}$ on Monday fell to $61\frac{3}{4}$ on Tuesday. How much did it fall?

(A) $2\frac{1}{8}$ points    (B) $2\frac{3}{8}$ points    (C) $2\frac{7}{8}$ points    (D) $3\frac{3}{8}$ points

(E) $3\frac{5}{8}$ points

# 8.  Size of Fractions

It is important to recognize which of several fractions is the smallest.

## A. FRACTIONS WITH THE SAME DENOMINATOR

Write

$a < b$    if $a$ is *smaller than* $b$

When this is the case, then $b$ is *larger than a*. This is written


$b > a$

For example,

$2 < 5$    (2 is smaller than 5)

and also,

$5 > 2$    (5 is larger than 2)

Observe that the symbols

$<, \quad >$

each point to the *smaller* number.

When two fractions have the *same* denominator, the one with the smaller numerator is the smaller fraction. Thus, if $a$, $b$, and $c$ are positive integers,

$$\frac{a}{b} < \frac{c}{b} \quad \text{if} \quad a < c$$

For example,

$$\frac{1}{4} < \frac{3}{4} \qquad \text{because} \qquad 1 < 3$$

**Example 1 ▶** Which of these fractions is the smallest?

(A) $\frac{5}{12}$      (B) $\frac{1}{12}$      (C) $\frac{7}{12}$      (D) $\frac{9}{12}$      (E) $\frac{13}{12}$

**Solution.** Each of these fractions has denominator 12. Among these, the smallest numerator is 1. Thus the smallest fraction is $\frac{1}{12}$. The correct choice is (B). ◀

*Try the exercises for Topic A on page 60.*

## B. TWO FRACTIONS WITH DIFFERENT DENOMINATORS

To compare the size of two (positive) fractions with *different* denominators, first find equivalent fractions with the same denominator. Then compare the numerators, as before.

It is often easiest to multiply the denominators to obtain a common denominator, even though this may not be the *least common denominator*.

**Example 2 ▶** Which is smaller? $\frac{3}{4}$ or $\frac{5}{6}$

**Solution.** Write equivalent fractions with denominator 24 (or 4 × 6).

$$\frac{3}{4} = \frac{6 \times 3}{6 \times 4} = \frac{18}{24}$$

$$\frac{5}{6} = \frac{4 \times 5}{4 \times 6} = \frac{20}{24}$$

Thus $\frac{3}{4} < \frac{5}{6}$    because    18 < 20.

Observe that you could also obtain this by "cross-multiplying":

$$\frac{3}{4} \diagup\!\!\!\!\diagdown \frac{5}{6}$$

$$3 \times 6 < 5 \times 4$$

and therefore

$$\frac{3}{4} < \frac{5}{6}$$ ◀

In general, for positive integers $a, b, c,$ and $d,$

$$\frac{a}{b} < \frac{c}{d} \quad \text{if} \quad a \times d < b \times c$$

*Try the exercises for Topic B on page 60.*

## C. FINDING THE SMALLEST FRACTION

**Example 3 ▶** Which of these fractions is the smallest?

(A) $\frac{3}{4}$  (B) $\frac{5}{8}$  (C) $\frac{7}{8}$  (D) $\frac{3}{5}$  (E) $\frac{5}{7}$

*Solution.* Compare (A) with (B); then compare the smaller of these with (C), and so on.

(A), (B): $\frac{5}{8} < \frac{3}{4}$  because  $5 \times 4 < 3 \times 8$  or  $20 < 24$

(A), (B), (C): $\frac{5}{8} < \frac{7}{8}$  because  $5 < 7$

(A), (B), (C), (D): $\frac{3}{5} < \frac{5}{8}$  because  $3 \times 8 < 5 \times 5$  or  $24 < 25$

(A), (B), (C), (D), (E): $\frac{3}{5} < \frac{5}{7}$  because  $3 \times 7 < 5 \times 5$  or  $21 < 25$

The smallest of these fractions is $\frac{3}{5}$. The correct choice is (D).  ◀

*Try the exercises for Topic C on page 61.*

## D. FINDING THE LARGEST FRACTION

**Example 4 ▶** Which of these fractions is the largest?

(A) $\frac{2}{7}$  (B) $\frac{3}{7}$  (C) $\frac{5}{7}$  (D) $\frac{3}{8}$  (E) $\frac{5}{8}$

*Solution.* Fractions (A), (B), and (C) all have denominator 7. Among these, clearly $\frac{5}{7}$ is the largest. Fractions (D) and (E) have denominator 8. The larger of fractions (D) and (E) is $\frac{5}{8}$. Finally,

$$\frac{5}{8} < \frac{5}{7} \quad \text{because} \quad 5 \times 7 < 8 \times 5 \quad \text{or} \quad 35 < 40$$

Observe that when two (positive) fractions have the same numerator, the *smaller fraction* corresponds to the *larger denominator*. Thus the *largest* of the five fractions is $\frac{5}{7}$ [choice (C)].  ◀

*Try the exercises for Topic D on page 62.*

## E. APPLICATIONS

**Example 5 ▶** Which is a better buy—a 2-ounce box of raisins for 25 cents or a 3-ounce box for 37 cents?

**Solution.** We want to increase the amount of raisins we get for each cent spent. Thus form the fractions

$$\frac{2 \text{ (ounces)}}{25 \text{ (cents)}} \quad \text{and} \quad \frac{3 \text{ (ounces)}}{37 \text{ (cents)}}$$

Compare $\frac{2}{25}$ with $\frac{3}{37}$. Clearly

$$\underbrace{2 \times 37}_{74} < \underbrace{3 \times 25}_{75}$$

Thus

$$\frac{2}{25} < \frac{3}{37}$$

Therefore the 3-ounce box for 37 cents is the better buy.    ◀

*Try the exercises for Topic E on page 62.*

## EXERCISES

**A.** *In each exercise, which fraction is the smallest? Draw a circle around the correct letter.*

1. (A) $\frac{2}{5}$      (B) $\frac{3}{5}$      (C) $\frac{4}{5}$

2. (A) $\frac{4}{9}$      (B) $\frac{2}{9}$      (C) $\frac{5}{9}$      (D) $\frac{8}{9}$

3. (A) $\frac{7}{10}$      (B) $\frac{9}{10}$      (C) $\frac{1}{10}$      (D) $\frac{3}{10}$

4. (A) $\frac{4}{13}$      (B) $\frac{7}{13}$      (C) $\frac{3}{13}$      (D) $\frac{11}{13}$      (E) $\frac{12}{13}$

5. (A) $\frac{4}{15}$      (B) $\frac{11}{15}$      (C) $\frac{7}{15}$      (D) $\frac{2}{15}$      (E) $\frac{13}{15}$

6. (A) $\frac{7}{20}$      (B) $\frac{9}{20}$      (C) $\frac{11}{20}$      (D) $\frac{3}{20}$      (E) $\frac{23}{20}$

**B.** *In each exercise, which is smaller?*

7. $\frac{1}{2}$ or $\frac{3}{5}$      *Answer:*          8. $\frac{3}{7}$ or $\frac{4}{9}$      *Answer:*

9. $\frac{2}{5}$ or $\frac{1}{3}$      *Answer:*          10. $\frac{3}{10}$ or $\frac{2}{7}$      *Answer:*

11. $\frac{7}{8}$ or $\frac{5}{6}$        *Answer:*          12. $\frac{7}{9}$ or $\frac{4}{5}$        *Answer:*

13. $\frac{9}{11}$ or $\frac{11}{13}$        *Answer:*          14. $\frac{3}{14}$ or $\frac{5}{13}$        *Answer:*

15. $\frac{9}{16}$ or $\frac{7}{13}$        *Answer:*          16. $\frac{10}{19}$ or $\frac{11}{24}$        *Answer:*

C. *In each exercise, which fraction is the smallest. Draw a circle around the correct letter.*

17.  (A) $\frac{1}{4}$       (B) $\frac{3}{4}$       (C) $\frac{1}{5}$       (D) $\frac{3}{5}$       (E) $\frac{4}{5}$

18.  (A) $\frac{7}{8}$       (B) $\frac{3}{8}$       (C) $\frac{1}{8}$       (D) $\frac{1}{9}$       (E) $\frac{7}{9}$

19.  (A) $\frac{1}{10}$       (B) $\frac{1}{9}$       (C) $\frac{1}{8}$       (D) $\frac{1}{7}$       (E) $\frac{1}{6}$

20.  (A) $\frac{3}{8}$       (B) $\frac{3}{10}$       (C) $\frac{3}{13}$       (D) $\frac{1}{9}$       (E) $\frac{1}{11}$

21.  (A) $\frac{2}{5}$       (B) $\frac{1}{3}$       (C) $\frac{3}{7}$       (D) $\frac{2}{9}$       (E) $\frac{2}{11}$

22.  (A) $\frac{7}{10}$       (B) $\frac{2}{3}$       (C) $\frac{3}{5}$       (D) $\frac{5}{8}$       (E) $\frac{8}{11}$

23.  (A) $\frac{4}{9}$       (B) $\frac{1}{2}$       (C) $\frac{2}{5}$       (D) $\frac{5}{11}$       (E) $\frac{3}{7}$

24.  (A) $\frac{9}{10}$       (B) $\frac{8}{9}$       (C) $\frac{9}{11}$       (D) $\frac{7}{8}$       (E) $\frac{19}{20}$

25.  (A) $\frac{1}{6}$       (B) $\frac{1}{3}$       (C) $\frac{2}{11}$       (D) $\frac{1}{10}$       (E) $\frac{2}{9}$

26.  (A) $\frac{3}{4}$       (B) $\frac{7}{10}$       (C) $\frac{4}{5}$       (D) $\frac{7}{9}$       (E) $\frac{5}{8}$

27.  (A) $\frac{1}{10}$       (B) $\frac{9}{100}$       (C) $\frac{91}{1000}$       (D) $\frac{3}{20}$       (E) $\frac{1}{5}$

28.  (A) $\frac{3}{5}$       (B) $\frac{5}{8}$       (C) $\frac{5}{9}$       (D) $\frac{7}{10}$       (E) $\frac{2}{3}$

29.  (A) $\frac{7}{6}$       (B) $\frac{5}{4}$       (C) $\frac{4}{3}$       (D) $\frac{6}{5}$       (E) $\frac{8}{7}$

30.  (A) $\frac{3}{10}$       (B) $\frac{2}{7}$       (C) $\frac{4}{9}$       (D) $\frac{3}{8}$       (E) $\frac{5}{12}$

31.  (A) $\frac{7}{10}$    (B) $\frac{4}{5}$    (C) $\frac{7}{9}$    (D) $\frac{7}{8}$    (E) $\frac{9}{11}$

32.  (A) $\frac{5}{12}$    (B) $\frac{7}{16}$    (C) $\frac{1}{2}$    (D) $\frac{2}{5}$    (E) $\frac{3}{8}$

33.  (A) $\frac{8}{11}$    (B) $\frac{3}{4}$    (C) $\frac{7}{9}$    (D) $\frac{5}{6}$    (E) $\frac{4}{5}$

34.  (A) $\frac{2}{9}$    (B) $\frac{1}{4}$    (C) $\frac{3}{8}$    (D) $\frac{2}{7}$    (E) $\frac{3}{11}$

35.  (A) $\frac{7}{20}$    (B) $\frac{1}{3}$    (C) $\frac{5}{11}$    (D) $\frac{5}{12}$    (E) $\frac{4}{15}$

36.  (A) $\frac{8}{13}$    (B) $\frac{7}{11}$    (C) $\frac{5}{7}$    (D) $\frac{13}{20}$    (E) $\frac{7}{12}$

**D.** *In each exercise, which fraction is the largest? Draw a circle around the correct letter.*

37.  (A) $\frac{1}{2}$    (B) $\frac{1}{5}$    (C) $\frac{1}{6}$    (D) $\frac{1}{7}$    (E) $\frac{1}{8}$

38.  (A) $\frac{7}{9}$    (B) $\frac{2}{9}$    (C) $\frac{5}{9}$    (D) $\frac{2}{3}$    (E) $\frac{1}{3}$

39.  (A) $\frac{5}{8}$    (B) $\frac{4}{7}$    (C) $\frac{3}{5}$    (D) $\frac{2}{3}$    (E) $\frac{7}{10}$

40.  (A) $\frac{2}{5}$    (B) $\frac{1}{3}$    (C) $\frac{4}{11}$    (D) $\frac{3}{8}$    (E) $\frac{5}{12}$

41.  (A) $\frac{7}{9}$    (B) $\frac{3}{4}$    (C) $\frac{8}{11}$    (D) $\frac{13}{16}$    (E) $\frac{7}{10}$

42.  (A) $\frac{1}{8}$    (B) $\frac{2}{11}$    (C) $\frac{3}{20}$    (D) $\frac{2}{15}$    (E) $\frac{1}{10}$

43.  (A) $\frac{5}{3}$    (B) $\frac{7}{5}$    (C) $\frac{5}{4}$    (D) $\frac{8}{7}$    (E) $\frac{11}{9}$

44.  (A) $\frac{7}{15}$    (B) $\frac{11}{20}$    (C) $\frac{1}{2}$    (D) $\frac{4}{7}$    (E) $\frac{6}{11}$

**E.**

45.  Which weighs more? $\frac{1}{2}$ pound of feathers or $\frac{7}{16}$ pound of lead?  *Answer:*

46.  Henry's portion is $\frac{3}{8}$ of the pie, Walter's is $\frac{2}{5}$. Who receives the larger portion?

   *Answer:*

**47.** One side of a kerchief measures $\frac{13}{20}$ of a meter. An adjacent side measures $\frac{7}{10}$ of a meter. Which side is longer?    *Answer:*

**48.** A salad dressing calls for $\frac{3}{4}$ cup of wine and $\frac{2}{3}$ cup of vinegar. Is there more wine or more vinegar in the dressing?    *Answer:*

**49.** Which is a better buy—a 6-ounce bar of soap for 75 cents or an 8-ounce bar for 92 cents?    *Answer:*

**50.** Which is a better buy—a 12-ounce can of string beans for 64 cents or a pound-can for a dollar?    *Answer:*

# 9.  Size of Decimals

Recall that the first decimal digit, that is, the digit that lies directly to the *right* of the decimal point, indicates the number of *tenths*; the second decimal digit, the number of *hundredths*; the third decimal digit, the number of *thousandths*. For example,

$$.7 = \frac{7}{10}, \qquad .43 = \frac{43}{100}, \qquad .043 = \frac{43}{1000}$$

$$\begin{array}{l}\text{tenths} \\ \text{hundredths} \\ \text{thousandths}\end{array}$$

(See pages 11 and 12 for a review of this material.)

## A.    COMPARING TENTHS

Clearly,

$$.7 < .9$$

because

$$\frac{7}{10} < \frac{9}{10}$$

**Example 1** ▶  Which of the following numbers is the smallest?

(A) .4          (B) .5          (C) .2          (D) .8          (E) .7

*Solution.* Each of these numbers has one decimal digit, and thus represents *tenths*. Among these, the smallest is .2 $\left( \text{or } \frac{2}{10} \right)$. Thus the correct choice is (C).                                                                    ◀

*Try the exercises for Topic A on page 68.*

## B. COMPARING HUNDREDTHS, THOUSANDTHS

Clearly,

$$.38 < .41$$

because

$$\frac{38}{100} < \frac{41}{100}$$

Thus to compare .$\boxed{3}$8 with .$\boxed{4}$1, first compare their tenths digits. Because $3 < 4$, it follows that

$$.38 < .41$$

Furthermore,

$$.43 < .47$$

because

$$\frac{43}{100} < \frac{47}{100}$$

Here both decimals, .43 and .47, have the same tenths digit, 4. Thus compare their hundredths digits. Because $3 < 7$, it follows that .4$\boxed{3}$ < .4$\boxed{7}$.

---

*To compare the size of two decimals:*

1. First compare their tenths digits.
2. If they are the same, compare their hundredths digits.
3. If they are also the same, compare their thousandths digits, and so on.

---

Thus

$$.\boxed{5}6 < .\boxed{6}3$$

because $5 < 6$;

$$.7\boxed{6}4 < .7\boxed{8}3$$

because $6 < 8$; and

$$.59\boxed{4}3 < .59\boxed{6}2$$

because $4 < 6$.

**Example 2 ▶**    Which of these numbers is the smallest?

(A) .27        (B) .29        (C) .31        (D) .25        (E) .35

*Solution.* First compare the tenths digits. Among these numbers, (A), (B), and (D) have the smaller tenths digits, 2. Now compare the hundredths digits of (A), (B), and (D). The smallest hundredths digit among these is 5, so that .25 is the smallest of the given numbers. Thus the correct choice is (D), or .25.    ◀

**Example 3 ▶**    Which of these numbers is the smallest?

(A) .614        (B) .579        (C) .597        (D) .609        (E) .578

*Solution.* First compare the tenths digits, then the hundredths digits, then the thousandths digits. Choices (B), (C), and (E) each have the smallest tenths digit, 5. *Among these*, (B) and (E) each have the smallest hundredths digit, 7. Between (B) and (E), choice (E) has the smaller thousandths digit, 8. Thus .578 is the smallest of these numbers. The correct choice is (E).    ◀

*Try the exercises for Topic B on page 68.*

## C. ADDING 0'S AT THE RIGHT

To compare .6 with .64, observe that

$$.6 = .60$$

and

$$.6\boxed{0} < .6\boxed{4}$$

Similarly, to compare .83 with .829, observe that

$$.83 = .830$$

and

$$.8\boxed{2}9 < .8\boxed{3}0$$

**Example 4 ▶**    Which of these numbers is the smallest?

(A) .54        (B) .541        (C) .504        (D) .5        (E) .55

**Solution.** Add 0's at the right, so that each of these has three decimal digits. Thus compare:

(A) .540     (B) .541     (C) .504     (D) .500     (E) .550

All have the same tenths digit, 5. Choices (C) and (D) each have the smallest hundredths digit, 0. Finally, between (C) and (D), choice (D) has the smaller thousandths digit, 0. Thus the smallest number is .500, or .5, so that the correct choice is (D).    ◀

*Try the exercises for Topic C on page 68.*

## D. MIXED DECIMALS

Clearly,

4.3 < 5.2

whereas

3.14 < 3.24

and

3.87 < 3.89

---

*To compare "mixed decimals":*

First compare their integral parts—then, if necessary, their tenths digits, then their hundredths digits, and so on.

---

**Example 5** ▶   Which is the smallest number?

(A) 2.03     (B) .032     (C) .023     (D) 0.23     (E) 0.03

**Solution.**  In (B) and (C) the integral part is understood to be 0. Thus choices (B), (C), (D), and (E) each have the smallest integral part, 0. Among these, (B), (C), and (E) each have the smallest tenths digit, 0. And among these, (C) has the smallest hundredths digit, 2. The smallest of these numbers is thus .023, which is choice (C).    ◀

*Try the exercises for Topic D on page 69.*

## E. LARGEST DECIMAL

**Example 6** ▶   Which is the largest number?

(A) .048     (B) .084     (C) .804     (D) .840     (E) .480

**Solution.**  The *largest* tenths digit is 8, in choices (C) and (D). Between these two, the larger hundredths digit is 4, in choice (D). Thus, the largest number is .840, that is, choice (D).    ◀

*Try the exercises for Topic E on page 69.*

**EXERCISES**

A. *In each exercise, which number is the smallest? Draw a circle around the correct letter.*

   1.  (A) .7       (B) .4       (C) .3       (D) .9       (E) .1

   2.  (A) .6       (B) .5       (C) .4       (D) .3       (E) .2

B. *In each exercise, which number is the smallest? Draw a circle around the correct letter.*

   3.  (A) .15    (B) .19    (C) .12    (D) .11    (E) .13

   4.  (A) .68    (B) .66    (C) .67    (D) .69    (E) .65

   5.  (A) .42    (B) .41    (C) .48    (D) .38    (E) .39

   6.  (A) .72    (B) .27    (C) .29    (D) .92    (E) .91

   7.  (A) .09    (B) .30    (C) .29    (D) .30    (E) .21

   8.  (A) .61    (B) .51    (C) .41    (D) .31    (E) .81

   9.  (A) .41    (B) .53    (C) .39    (D) .43    (E) .51

  10.  (A) .08    (B) .09    (C) .11    (D) .18    (E) .10

  11.  (A) .451   (B) .458   (C) .455   (D) .459   (E) .453

  12.  (A) .308   (B) .302   (C) .306   (D) .304   (E) .307

  13.  (A) .753   (B) .575   (C) .759   (D) .751   (E) .579

  14.  (A) .039   (B) .093   (C) .309   (D) .390   (E) .309

  15.  (A) .111   (B) .101   (C) .110   (D) .100   (E) .102

  16.  (A) .083   (B) .008   (C) .038   (D) .003   (E) .033

  17.  (A) .148   (B) .139   (C) .149   (D) .137   (E) .136

  18.  (A) .707   (B) .770   (C) .777   (D) .077   (E) .070

C. *In each exercise, which number is the smallest?*

  19.  (A) .707   (B) .703   (C) .77    (D) .737   (E) .730

  20.  (A) .04    (B) .039   (C) .03    (D) .304   (E) .43

  21.  (A) .614   (B) .6014  (C) .6104  (D) .61    (E) .601

  22.  (A) .899   (B) .89    (C) .9     (D) .891   (E) .8901

23. (A) .704        (B) .740        (C) .709        (D) .741        (E) .7

24. (A) .321        (B) .3201       (C) .3210       (D) .33         (E) .3202

25. (A) .195        (B) .159        (C) .19         (D) .155        (E) .1501

26. (A) .1401       (B) .1104       (C) .1004       (D) .101        (E) .11

**D.** *In each exercise, which number is the smallest? Draw a circle around the correct letter.*

27. (A) 1.07        (B) 2.07        (C) 3.07        (D) .07         (E) 1.09

28. (A) 4.2         (B) 2.4         (C) 2.24        (D) 4.02        (E) 2.04

29. (A) 12.9        (B) 10.9        (C) 10.09       (D) 12.009      (E) 10.009

30. (A) 6.2         (B) 2.6         (C) 2.606       (D) 2.06        (E) 2.066

31. (A) 3.4         (B) 3.389       (C) 3.408       (D) 3.409       (E) 3.4089

32. (A) 1.85        (B) 1.08        (C) 0.85        (D) .088        (E) .009

33. (A) .004        (B) .04         (C) 1.004       (D) .045        (E) .4

34. (A) .6          (B) .06         (C) .0606       (D) .066        (E) 6.006

35. (A) 3.079       (B) .0379       (C) .0039       (D) .0309       (E) 3.009

36. (A) .0030       (B) .03         (C) .029        (D) .031        (E) .0309

37. (A) 2.02        (B) 2.20        (C) .022        (D) .0202       (E) 2.022

38. (A) .848        (B) .884        (C) .088        (D) .0884       (E) .0808

**E.** *In each exercise, which number is the largest? Draw a circle around the correct letter.*

39. (A) .73         (B) .78         (C) .87         (D) .83         (E) .79

40. (A) .305        (B) .503        (C) .055        (D) .053        (E) .505

41. (A) 1.2         (B) 1.21        (C) 1.02        (D) 2.01        (E) 2.11

42. (A) .022        (B) .032        (C) .201        (D) .2001       (E) .2031

43. (A) 6.1         (B) 4.9         (C) 5.2         (D) 5.9         (E) 6.01

44. (A) 4.08        (B) .048        (C) .008        (D) .084        (E) 8.04

45. (A) 6.01        (B) .061        (C) .6011       (D) 6.009       (E) 6.0101

46. (A) .099        (B) .0909       (C) .0991       (D) .0919       (E) .0990

# 10.   Adding and Subtracting Decimals

## A. ADDING DECIMALS

> *To add or subtract decimals:*
>
> 1.  Line up the decimal digits.
> 2.  Add or subtract in columns.

Thus

$$\begin{array}{r} .31 \\ + .48 \\ \hline .79 \end{array}$$    because    $\dfrac{31}{100} + \dfrac{48}{100} = \dfrac{79}{100}$

**Example 1 ▶**   Add.    $10.45 + 11.04 + 8.82$

*Solution.*

$$\begin{array}{r} 10.45 \\ 11.04 \\ 8.82 \\ \hline 30.31 \end{array}$$

The sum is 30.31.                                                    ◀

You may want to add 0's to the right of the last decimal digit, although this is not necessary.

**Example 2 ▶**   Add.    $14.42 + 10.044 + 8$

*Solution.*

$$\begin{array}{r} 14.\ 42\boxed{0} \\ 10.\ 04\ 4 \\ 8.\boxed{00\ 0} \\ \hline 32.\ 46\ 4 \end{array}$$

The sum is 32.464.    ◀

*Try the exercises for Topic A on this page.*

## B. SUBTRACTING DECIMALS

**Example 3** ▶    Subtract.    49.3 – 4.84

> *Solution.*  First line up the decimal points. Change 49.3 to 49.3 $\boxed{0}$. Then subtract in columns.
>
> ```
>      8 12 10                    13 10
>     49.3̸ 0̸        or        49.3̸ 0̸
>    -  4.8 4                    5 9
>    ───────                  -  4.8 4
>     44.4 6                   ───────
>                               44.4 6
> ```

The difference is 44.46.    ◀

*Try the exercises for Topic B on page 72.*

## C.    A FURTHER EXAMPLE

**Example 4** ▶    14.2 – 2.04 =

> *Solution.*
> ```
>        1 10                       10
>     14.2̸ 0̸        or        14.2 0̸
>    -  2.0 4                    1
>    ───────                  -  2.0̸ 4
>     12.1 6                   ───────
>                               12.1 6
> ```

*Try the exercises for Topic C on page 72.*    ◀

## EXERCISES

**A.**    *Add.*

1.  7.2 + 3.8 + 9.5 =

2.  6.4 + 3.6 + 8.4 =

3.  6.82 + 4.85 + 3.06 =

4.  10.41 + 8.09 + 7.96 =

5.  7.71 + 12.9 + 6 =

6.  8.03 + 3.1 + 16 =

7.  19.92 + 9.908 + 10 =

8.  27.2 + 2.04 + 12 =

9.  1.04 + 104 + 10.4 =

10.  6.08 + 60.8 + 68 =

11.  12.01 + 1.201 + 120 =

12.  5.99 + 15.09 + 1.909 =

13.  109.9 + 10.99 + 1.099 =

14.  19.09 + 1.099 + 109 =

**B.** *Subtract.*

15.  12.7 − 11.3 =

16.  16.4 − 8.4 =

17.  42.3 − 31.4 =

18.  13.7 − 7.9 =

19.  12.1 − 6.06 =

20.  10.03 − 8.09 =

21.  43.8 − 2.07 =

22.  4.06 − .093 =

23.  1.092 − .909 =

24.  52.05 − 8.559 =

25.  25.8 − 1.29 =

26.  41.7 − 14.03 =

27.  54.6 − 9.83 =

28.  101.4 − 1.014 =

**C.**  *Add or subtract, as indicated.*

29.  2.05 + 5.2 + 4.52 =

30.  7.83 + 3.87 + 5.087 =

31.  6.03 + 16.3 + 11 =

32.  15.83 + 5.805 + 12 =

33.  10.09 + 9.021 + 8 =

34.  94.01 + 4.094 + 19 =

35.  9.093 + 90.39 + 18 =

36.  9.08 − 6.12 =

37.  10 − 4.09 =

38.  24.1 − 4.02 =

39.  19.03 − 2.89 =

40.  39.6 − 9.88 =

41.  17.8 − 2.036 =

42.  41.9 − 3.28 =

# 11.  Multiplying and Dividing Decimals

There are many important applications of multiplication and division of decimals. For example, you may want to find the cost of a certain item at a fixed cost per item. The cost is in dollars and cents, so that you are essentially multiplying by a decimal. Also, division of decimals helps you on certain percent problems (Section 14).

## A. MULTIPLYING DECIMALS

> When one or both factors of a product are decimals, count the total number of decimal digits in the factors in order to place the decimal point in the product.

For example,

$$.7 \quad \times \quad .11 \quad = \frac{7}{10} \times \frac{11}{100} = \frac{77}{1000} = .\boxed{0}\,77$$

1 decimal digit     2 decimal digits     3 decimal digits

$1\boxed{0} \qquad 1\boxed{00} \qquad 1\boxed{000}$

Note that you must add a 0 *immediately to the right of the decimal point* in order to obtain 3 decimal digits in the product.

**Example 1** ▶ Multiply.    $18 \times .25$

*Solution.*

$$
\begin{array}{r}
18 \longleftarrow \text{0 decimal digits} \\
\times\ .25 \longleftarrow + \text{2 decimal digits} \\
\hline
90 \\
3\,6 \\
\hline
4.50 \longleftarrow \text{2 decimal digits}
\end{array}
$$

The final 0 counts as a decimal digit, even though it can be dropped. Thus

$$18 \times .25 = 4.50 = 4.5$$

◀

**Example 2** ▶  Multiply.    $.38 \times 6.2$

**Solution.**

$$
\begin{array}{r}
.3\,8 \quad\longleftarrow\; \text{2 decimal digits} \\
\underline{\times\;\; 6.2} \quad\longleftarrow\; \underline{\text{+ 1 decimal digit}} \\
7\,6 \phantom{0}\\
\underline{2\,28 \phantom{00}}\\
2.35\,6 \quad\longleftarrow\; \text{3 decimal digits}
\end{array}
$$

◀

*Try the exercises for Topic A on page 76.*

## B.  COST PROBLEMS

**Example 3** ▶  Find the cost of 17 gallons of gas at \$1.28 per gallon.

**Solution.**  Each gallon of gas costs the same amount. Thus you *multiply* the number of gallons, 17, by the fixed cost per gallon, \$1.28.

$$
\begin{array}{r}
\$\;1.28 \quad\longleftarrow\; \text{2 decimal digits} \\
\underline{\times\;\; 17} \quad\longleftarrow\; \underline{\text{+ 0 decimal digits}} \\
8\,96 \phantom{0}\\
\underline{12\,8 \phantom{00}}\\
\$21.76 \quad\longleftarrow\; \text{2 decimal digits}
\end{array}
$$

The cost of 17 gallons is \$21.76.

◀

*Try the exercises for Topic B on page 76.*

## C.  DIVIDING DECIMALS

When dividing by *ten*ths, first multiply dividend and divisor by 10, so that the divisor will then be an integer. For example,

$$\frac{24}{.6} = \frac{24 \times 10}{.6 \times 10} = \frac{240}{6} = 40$$

When dividing by *hundred*ths, first multiply dividend and divisor by 100, so that the divisor will then be an integer. Thus

$$\frac{45}{.03} = \frac{45 \times 100}{.03 \times 100} = \frac{4500}{3} = 1500$$

**Example 4** ▶  Divide 139.4 by .17.

**Solution.**

$$\frac{139.4}{.17} = \frac{139.4 \times 100}{.17 \times 100} = \frac{13,940}{17}$$

Note that to multiply

$$139.4 \times 100$$

move the decimal point 2 places to the right. In order to do so, insert a 0 to the right of the 4.

$$139.4 \times 100 = 13,94\boxed{0}.$$

Now divide 13,940 by 17.

```
            820
   17⟌13 940
       13 6
          34
          34
           0
```

Thus the quotient is 820.                                                    ◀

*Try the exercises for Topic C on page 76.*

## D.  COST PROBLEMS

**Example 5** ▶  Cans of soda cost $.40 each. How many can be bought for $12?

**Solution.**  Divide the total cost, $12, by the cost per can, $.40. Note that .40 = .4.

*Multiply numerator and denominator by* 10.

$$\frac{\$12}{\$.40} = \frac{12 \times 10}{.4 \times 10} = \frac{120}{4} = 30$$

Thus 30 cans of soda at $.40 each can be bought for $12.                      ◀

*Try the exercises for Topic D on page 77.*

## E.  MULTIPLE CHOICE

**Example 6** ▶  48 bagels at $.24 per bagel cost a total of

(A) $.20        (B) $2.00        (C) $200        (D) $1152        (E) $11.52

**Solution.**  You are given the cost *per* bagel, $.24, and want to find the cost of 48 bagels. Thus multiply $.24 by 48.

```
   $   .24  ◀──── 2 decimal digits
   ×  48    ◀──── + 0 decimal digits
     1 92
     9 6
   $11.52   ◀──── 2 decimal digits
```

The cost is $11.52. The correct choice is (E).    ◄

*Try the exercises for Topic E on page 77.*

## EXERCISES

### A. *Multiply.*

1.  .3 × .5    *Answer:*        2.  .4 × .2    *Answer:*

3.  .14 × .8    *Answer:*       4.  .12 × .3    *Answer:*

5.  .16 × .5    *Answer:*       6.  .48 × .07    *Answer:*

7.  .92 × .29    *Answer:*      8.  22 × .5    *Answer:*

9.  48 × .05    *Answer:*       10.  36 × 1.2    *Answer:*

11.  280 × .41    *Answer:*     12.  38.9 × 1.02    *Answer:*

13.  503 × 2.15    *Answer:*    14.  .395 × 92.1    *Answer:*

### B.

15.  Find the cost of 12 ice cream cones at $.65 per cone.    *Answer:*

16.  Find the cost of 8 melons at $1.15 per melon.    *Answer:*

17.  A book salesman sells 6 copies of a book priced at $10.25 per copy. What is the total sale price?    *Answer:*

18.  A can of coffee sells for $2.95. What is the price of 7 cans of this coffee?    *Answer:*

19.  Apples sell for $.68 per pound. What is the price of 5 pounds of apples?    *Answer:*

20.  How much do 6 quarts of milk cost at $.71 per quart?    *Answer:*

21.  Each page of a book is .003 inches thick. How thick is a 500-page book?    *Answer:*

22.  A roll of wallpaper costs $9.75. If a room requires 26 rolls of this wallpaper, what is the cost of wallpaper for this room?    *Answer:*

23.  What is the cost of 47 gallons of paint at $8.95 per gallon?    *Answer:*

24.  A woman earns $6.25 per hour of work. How much does she earn for 36 hours of work?    *Answer:*

### C. *Divide.*

25.  25 ÷ .5    *Answer:*       26.  120 ÷ .6    *Answer:*

27.  32 ÷ .04    *Answer:*      28.  910 ÷ .07    *Answer:*

29.  $4800 \div .12$      *Answer:*          30.  $121 \div .11$      *Answer:*

31.  $2.6 \div .13$      *Answer:*          32.  $840 \div .12$      *Answer:*

33.  $13.5 \div 4.5$      *Answer:*          34.  $14.4 \div .48$      *Answer:*

35.  $5.2 \div .13$      *Answer:*          36.  $10.24 \div .032$      *Answer:*

**D.**

37.  Pencils cost $.07 each. How many can be bought for $1.26?      *Answer:*

38.  Oranges cost $.30 each. How many can be bought for $5.10?      *Answer:*

39.  Erasers sell for $.60 each. How many can you buy for $14.40?
     *Answer:*

40.  Grapes sell for $1.60 per pound. How many pounds can you buy for $11.20?
     *Answer:*

41.  Sugar sells for $.75 per pound. How many pounds can you buy for $11.25?
     *Answer:*

42.  A man buys a car for $10,560. If he pays $440 in equal monthly installments,
     for how many months will the payments last?      *Answer:*

43.  A book sells for $15. One week, total sales for the book are $1080. How many
     copies of the book are sold that week?      *Answer:*

44.  A woman pays $21.60 for a tank of gas. If gas sells for $1.20 per gallon, how many
     gallons does she buy?      *Answer:*

**E.**  *Draw a circle around the correct letter.*

45.  $.12 \times .09 =$

     (A)  108          (B)  1.08          (C)  10.8          (D)  .108          (E)  .0108

46.  $1.4 \times .08 =$

     (A)  .112          (B)  .0112          (C)  .001 12          (D)  175          (E)  17.5

47.  $42.1 \times .21 =$

     (A)  2          (B)  20          (C)  200          (D)  8.841          (E)  88.2

48.  $60 \div .05 =$

     (A)  3          (B)  .03          (C)  12          (D)  120          (E)  1200

49.  $124 \div 3.1 =$

     (A)  .04          (B)  .4          (C)  4          (D)  40          (E)  400

**50.** $9.6 \div .016 =$

(A) 600        (B) 60        (C) 6        (D) .6        (E) .06

**51.** 32 cans of peas at $.45 per can cost

(A) $14.40     (B) $13.40     (C) $2.88     (D) $129.60     (E) $24.40

**52.** If notebooks cost $.85 apiece, how many can you buy for $16.15?

(A) 17        (B) 18        (C) 19        (D) 20        (E) 21

**53.** If a motorist uses 22 gallons of gasoline at $1.35 per gallon, what is the total price she pays for gasoline?

(A) $27.90     (B) $29.70     (C) $30.70     (D) $39.70     (E) $27.70

**54.** If each booklet is .18 inch thick and a pile of these booklets is 4.68 inches high, how many booklets are there in the pile?

(A) 260       (B) 2.60       (C) 216       (D) 16       (E) 26

---

# 12.  Cost and Profit

Here we consider how to determine the cost of an item or the profit made on it.

## A.  COST

**Example 1** ▶ Find the total cost of 5 pounds of potatoes at $.30 per pound and 3 pounds of onions at $.35 per pound.

(A) $1.50     (B) $1.05     (C) $2.55     (D) $25.55     (E) $.45

*Solution.* In each case, multiply the cost per pound by the number of pounds.

Potatoes:    $.30 × 5 = $1.50

Onions:    $.35 × 3 = $1.05

Now add the cost of potatoes and the cost of onions to find the total cost.

$1.50
+ $1.05
$2.55

The total cost is $2.55. The correct choice is (C).    ◀

**Example 2 ▶** A parking lot charges $3 for the first hour and $2 for each additional hour. What is the cost of parking a car there for 6 hours?

*Solution.* The cost for the first hour is $3. The car is parked there for 6 hours in all, so that there are 6 – 1, or 5, additional hours. For each additional hour the cost is $2. The *total* cost is then

$$\$3 + (6 - 1) \times \$2 = \$3 + (5 \times \$2)$$
$$= \$3 + \$10$$
$$= \$13$$    ◀

*Try the exercises for Topic A on page 80.*

## B. PROFIT

When an item is sold, the **profit** made equals the **total sales** minus the **costs**.

Profit = Total Sales – Costs

**Example 3 ▶** A grocer buys a carton of 18 boxes of cornflakes for $10.50. He sells each box for 85¢. Find his profit for selling all 18 boxes.

(A) $4.80    (B) $10.50    (C) $15.30    (D) $3.60    (E) $3.80

*Solution.* First, to find the *total sales*, multiply 18, the number of boxes sold, by 85¢, the price per box. Note that

85¢ = $.85

$   .85 ◄——— Price per item
X  18 ◄——— Number of items
6 80
8 5
$15.30 ◄——— Total Sales

Next, the *cost* to the grocer is $10.50. To find his *profit*, use

Profit = Total Sales – Costs
   =   $15.30   – $10.50

$15.30 ⟵——— Total Sale
– $10.50 ⟵——— Costs
$ 4.80 ⟵——— Profit

His profit is $4.80. The correct choice is (A).    ◀

**Example 4 ▶** A drama group sells 584 tickets to a play at $5 each. The group pays $750 to rent an auditorium and $325 in other expenses. Find the profit.

*Solution.* To find the *total sales*, multiply 584, the number of tickets sold, by $5, the price per ticket.

584 × $5 = $2920 ⟵—Total Sales

To find the *costs*, add $750, for renting the auditorium, and $325, for other expenses.

$ 750
+ $ 325
$1075 ⟵——— Costs

Profit = Total Sales – Costs
   =   $2920   – $1075

$2920 ⟵——— Total Sales
– $1075 ⟵——— Costs
$1845 ⟵——— Profit

The profit is $1845.    ◀

*Try the exercises for Topic B on page 81.*

## EXERCISES

A.   *In Exercises 1-6, draw a circle around the correct letter.*

1.  The cost to a store of 92 notebooks at 48¢ per notebook is

(A) $33.16    (B) $34.16    (C) $43.16    (D) $44.06    (E) $44.16

2.  Pens cost 25¢ each and erasers cost 35¢ each. The cost of 36 pens and 18 erasers is

(A) $15    (B) $15.30    (C) $6.30    (D) $32.40    (E) $17.10

3.  A bookstore orders 27 copies of a history book at $8 per copy and 40 copies of an atlas at $9.50 per copy. The total cost to the bookstore is

(A) $596    (B) $380    (C) $216    (D) $1172.50    (E) $117.25

4. A grocer orders 12 cartons of detergent at $27 per carton, 9 cartons of tissues at $11 per carton, and 15 cartons of paper towels at $13 per carton. The total cost to the grocer is

(A) $423    (B) $618    (C) $294    (D) $324    (E) $519

5. A moving company charges $95 for the first hour of work and $75 for each additional hour. The cost of an 8-hour job is

(A) $835    (B) $845    (C) $695    (D) $620    (E) $740

6. To mail a (first-class) letter in 1983 costs 20¢ for the first ounce and 17¢ for each additional ounce. The cost of mailing a 10-ounce letter is

(A) $2.00    (B) $1.70    (C) $1.73    (D) $1.90    (E) $1.97

7. To rent an auditorium costs $250 per hour for the first two hours and $150 for each additional hour. The cost to rent that auditorium for 5 hours is _____

8. A duplicating center charges 8 cents per copy for the first 5 copies of the same page, 6 cents per copy for the next 10 copies, and 5 cents per copy for each additional copy. The cost of making 100 copies of a page is _____

9. In Exercise 8, the cost of making 50 copies of one page and 20 copies of another page is _____

10. In Exercise 8, the cost of making 10 copies of one page, 25 copies of a second page, and 45 copies of a third page is _____

B. *In Exercises 11-16, draw a circle around the correct letter.*

11. A grocer buys a carton of 24 boxes of crackers for $9.60. She sells each box for 55 cents. Her profit for selling all 24 boxes is

(A) $22.80    (B) $13.20    (C) $3.60    (D) $3.20    (E) $4.60

12. A store buys a box of 32 scarves for $189. It sells each scarf for $9. The profit for selling all of the scarves is

(A) $278    (B) $288    (C) $199    (D) $99    (E) $109

13. A man bought 6 melons at $1.35 each. How much change did he receive from a $10 bill?

(A) $8.00    (B) $8.10    (C) $8.65    (D) $1.90    (E) $2.90

14. A woman bought 22 cans of soup at 65¢ each. How much change did she receive from a $20 bill?

(A) $14.30    (B) $6.70    (C) $5.70    (D) $.57    (E) $7.00

15. Tickets to a concert cost $9 each. If 880 tickets are sold and the expenses of running the concert are $6400, the profit is

(A) $7920    (B) $14,320    (C) $1520    (D) $5520    (E) $820

16. A druggist buys 48 tubes of toothpaste at $.53 each. She sells them for $.79 each. Her total profit is

    (A) $.26        (B) $12.48        (C) $124.80        (D) $25.44        (E) $37.92

17. A department store pays $1450 for a shipment of 35 coats. It sells them all for $79 apiece. The profit is _____

18. At a football game 512 tickets are sold for $6.50 each. It costs $2200 to rent the stadium. If other expenses amount to $750, the profit is _____

19. A theater charges $6 for an orchestra seat and $4 for a balcony seat. It sells 330 orchestra seats and 252 balcony seats. If expenses are $2400, the theater's profit is
_____

20. A supermarket buys two dozen loaves of bread for 60¢ per loaf. It sells 19 of these loaves for 87¢ each. If the other loaves get stale and are not sold, the profit is
_____

21. A store buys 60 umbrellas for $4.50 each. It sells 35 of the umbrellas for $8.00 each and it sells the remaining umbrellas at a reduced price of $5.95 each. The store's profit is _____

22. 550 people pay $12 each to see a play. The theater company pays the performers a total of $2400. It pays $1750 to rent the theater and $1225 in other expenses. Its profit is _____

# 13.  Fractions, Decimals, and Percent

What is the relationship between fractions, decimals, and percent? How do you change a fraction, such as $\frac{6}{13}$, to a decimal, rounded to the nearest hundredth or thousandth? And how do you express a percent as a fraction or decimal?

## A. FRACTIONS TO DECIMALS

To change a fraction to a decimal, divide the numerator by the denominator.

**Example 1 ▶**  Change $\frac{5}{8}$ to a decimal.

*Solution.* Divide the numerator, 5, by the denominator, 8. First write 5 as 5.000. (Add as many 0's as necessary to the right of the decimal point.)

$$
\begin{array}{r}
.625 \\
8\overline{)5.000}
\end{array}
$$ ◄—— Line up the decimal points.

Thus

$$\frac{5}{8} = .625$$ ◄

**Example 2 ▶**  Change $\frac{6}{13}$ to a decimal rounded to the nearest thousandth.

*Solution.* Divide the numerator, 6, by the denominator, 13.

$$
\begin{array}{r}
.4615\ldots \\
13\overline{)6.0000} \\
5\,2 \\
\hline
80 \\
78 \\
\hline
20 \\
13 \\
\hline
70 \\
65 \\
\hline
\end{array}
$$

You are asked to express this to three decimal digits (thousandths). When rounding to *three* decimal digits, if the *fourth* decimal digit is 5 or more, add 1 to the *third* decimal digit. Because the fourth decimal digit is 5 here, change the third decimal digit from 1 to 2, and express the final result as .462. ◄

*Try the exercises for Topic A on page 85.*

## B. DECIMALS TO FRACTIONS

To change from a decimal to a fraction, rewrite the decimal as a fraction. Then reduce to lowest terms.

**Example 3 ▶**  Change .075 to a fraction in lowest terms.

*Solution.*

$$.075 = \frac{75}{1000}$$    Divide numerator and denominator by 25.

$$= \frac{3}{40}$$ ◄

*Try the exercises for Topic B on page 86.*

## C. PERCENT TO FRACTIONS

**Percent** means *hundredths*. For example,

$$57\% = \frac{57}{100}$$

To express a percent as a fraction:

1. Remove the percent symbol, %.
2. Divide the resulting number by 100.
3. Reduce to lowest terms, if necessary.

**Example 4 ▶**    What is 85% expressed as a fraction?

*Solution.*

$$85\% = \frac{85}{100} \qquad \text{Divide numerator and denominator by 5.}$$

$$= \frac{17}{20}$$    ◀

*Try the exercises for Topic C on page 86.*

## D. PERCENT TO DECIMALS

To convert a percent to a decimal:

1. If there is no decimal point, insert one to the left of the percent symbol.
2. Remove the percent symbol.
3. Move the decimal point two digits to the left.

For example,

$$39\% = 39.\% = .39$$

and

$$8\% = 8.\% = .08$$

Note that when converting 8% to .08, you must insert a $\boxed{0}$ to the left of the 8, so that you can move the decimal point two digits to the left.

**Example 5 ▶**  Change 160% to a decimal.

> ***Solution.***
>
> $$160\% = 160.\% = 1.\underset{\smile}{60} = 1.6$$
>
> Note that the 0 at the far right in 1.60 can be dropped. Thus
>
> $$160\% = 1.6$$ ◀

*Try the exercises for Topic D on page 86.*

## E. MULTIPLE CHOICE

**Example 6 ▶**  Change $\frac{11}{17}$ to a decimal rounded to the nearest hundredth.

> (A) .647    (B) .64    (C) .65    (D) .6    (E) .60
>
> ***Solution.***  Divide the numerator, 11, by the denominator, 17.
>
> ```
>        .647
> 17 |11.000
>     10 2
>     ─────
>        80
>        68
>     ─────
>       120
>       119
>     ─────
> ```
>
> You are asked to express this to two decimal digits (hundredths). Because the third decimal digit is 7, which is 5 or more, change the second decimal digit from 4 to 5, and express the final result as .65. The correct choice is (C).    ◀

*Try the exercises for Topic E on page 87.*

## EXERCISES

**A.** *Change each fraction to a decimal.*

1. $\frac{4}{5}$     *Answer:*        2. $\frac{3}{4}$     *Answer:*

3. $\frac{5}{2}$     *Answer:*        4. $\frac{3}{8}$     *Answer:*

5. $\frac{7}{20}$     *Answer:*        6. $\frac{1}{25}$     *Answer:*

7. $\frac{7}{25}$     *Answer:*        8. $\frac{3}{50}$     *Answer:*

9. $\frac{1}{200}$     *Answer:*        10. $\frac{21}{40}$     *Answer:*

*Change each fraction to a decimal rounded to the nearest hundredth.*

11. $\frac{2}{3}$            *Answer:*            12. $\frac{4}{9}$            *Answer:*

13. $\frac{1}{6}$            *Answer:*            14. $\frac{3}{11}$            *Answer:*

15. $\frac{4}{13}$            *Answer:*            16. $\frac{7}{17}$            *Answer:*

*Change each fraction to a decimal rounded to the nearest thousandth.*

17. $\frac{5}{11}$            *Answer:*            18. $\frac{4}{7}$            *Answer:*

19. $\frac{8}{13}$            *Answer:*            20. $\frac{3}{17}$            *Answer:*

21. $\frac{12}{13}$            *Answer:*            22. $\frac{10}{19}$            *Answer:*

**B.**  *Change each decimal to a fraction in lowest terms.*

23.  .24            *Answer:*            24.  .35            *Answer:*

25.  .02            *Answer:*            26.  .08            *Answer:*

27.  .625            *Answer:*            28.  .035            *Answer:*

29.  1.2            *Answer:*            30.  1.25            *Answer:*

**C.**  *Express each percent as a fraction in lowest terms.*

31.  50%            *Answer:*            32.  40%            *Answer:*

33.  75%            *Answer:*            34.  35%            *Answer:*

35.  22%            *Answer:*            36.  36%            *Answer:*

37.  150%            *Answer:*            38.  175%            *Answer:*

39.  48%            *Answer:*            40.  64%            *Answer:*

**D.**  *Change each percent to a decimal.*

41.  53%            *Answer:*            42.  14%            *Answer:*

43.  40%            *Answer:*            44.  90%            *Answer:*

45.  2%            *Answer:*            46.  150%            *Answer:*

**47.** 225%          *Answer:*          **48.** 12.5%          *Answer:*

**E.** *Draw a circle around the correct letter.*

**49.** What is 15% expressed as a fraction?

(A)  15          (B)  1500          (C)  $\frac{3}{20}$          (D)  $\frac{3}{10}$          (E)  $1\frac{3}{20}$

**50.** What is 44% expressed as a fraction?

(A)  $\frac{4}{9}$          (B)  $\frac{4}{11}$          (C)  $\frac{44}{99}$          (D)  $\frac{11}{20}$          (E)  $\frac{11}{25}$

**51.** Change $\frac{1}{8}$ to a decimal.

(A) .18          (B) .1          (C) .12          (D) .125          (E) .1025

**52.** Change $\frac{7}{50}$ to a decimal.

(A) .7          (B) .07          (C) .14          (D) .014          (E) 7.14

**53.** Change $\frac{2}{7}$ to a decimal rounded to the nearest hundredth.

(A) .22          (B) .28          (C) .29          (D) .285          (E) .286

**54.** Change $\frac{3}{13}$ to a decimal rounded to the nearest thousandth.

(A) .23          (B) .230          (C) .2307          (D) .231          (E) 4.33

**55.** Change $\frac{1}{17}$ to a decimal rounded to the nearest hundredth.

(A) .58          (B) .59          (C) .058          (D) .05          (E) .06

**56.** Change $\frac{12}{17}$ to a decimal rounded to the nearest thousandth.

(A) .705          (B) .7058          (C) .706          (D) .715          (E) .7059

**57.** Change $\frac{3}{19}$ to a decimal rounded to the nearest hundredth.

(A) .15          (B) .157          (C) .16          (D) .158          (E) .17

**58.** Change $\frac{6}{7}$ to a decimal rounded to the nearest thousandth.

(A) .857          (B) .858          (C) .86          (D) .8571          (E) .8572

**59.** Change $\frac{5}{12}$ to a decimal rounded to the nearest thousandth.

    (A) .416      (B) .426      (C) .417      (D) .418      (E) .4167

**60.** Change $\frac{11}{14}$ to a decimal rounded to the nearest thousandth.

    (A) .78      (B) .785      (C) .7857      (D) .79      (E) .786

# 14.  Percent Problems

Questions involving percent arise frequently in stated problems. For example, you may be asked:

        What is 20% of 80?       (*Topic A*)

or

        30 is 20% of what number?     (*Topic C*)

Other percent problems will be discussed in Section 15.

**A. 20% OF 80**

Suppose you are asked

        What is 20% of 80?

Here, the word "of" indicates multiplication, so that what you want to find is

$$20\% \times 80$$

Change 20% to a decimal

$$20\% = .20 = .2$$

and then calculate

$$.2 \times 80 = 16.0 = 16$$

Thus 20% of 80 is 16.

**Example 1 ▶** What is 60% of 75?

(A) 60          (B) 50          (C) 45          (D) 4.5          (E) 125

*Solution.* You want

$$60\% \times 75$$

Change 60% to a decimal

$$60\% = .60 = .6$$

and then calculate

$$.6 \times 75 = 45.0 = 45$$

Thus 60% of 75 is 45. The correct choice is (C).          ◀

*Try the exercises for Topic A on page 93.*


## B. EQUATIONS

In order to solve some percent problems, it is best to set up an *equation*. Before considering an equation with percent, here is a simple illustration. Suppose you are told:

Four times a number is equal to 12

and asked to find this number. To do so, let *n* stand for this *unknown* number. Then translate the problem as follows:

Four times a number is equal to 12
4        ×        *n*        =        12

This last statement is known as an *equation*. An **equation** is a statement of *equality*.

An equation is like a balanced scale. In order to preserve the balance, if you change one side, you must do the same thing to the other side. (See the figure to the right.)

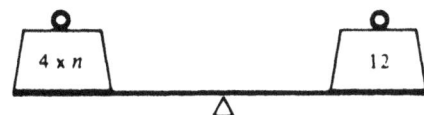

Because 4 *multiplies* n on the left side, in order to isolate the unknown, *divide* the left side by 4. But now the scale is unbalanced. (See the figure to the right.)

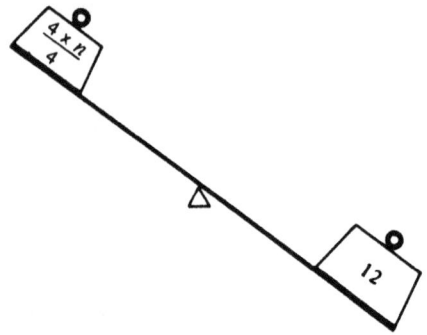

What you do to the left side, also do to the right side. Thus divide both sides by 4 to preserve the balance. (See the figure to the right.)

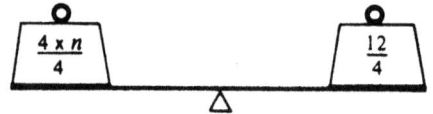

$$4 \times n = 12 \qquad \text{Divide both sides by 4.}$$

$$\frac{4 \times n}{4} = \frac{12}{4}$$

$$\frac{\overset{1}{\cancel{4} \times n}}{\underset{1}{\cancel{4}}} = \frac{\overset{3}{\cancel{12}}}{\underset{1}{\cancel{4}}}$$

$$n = 3$$

Thus the number n is now known to be 3. (Note that $4 \times 3 = 12$.)

**Example 2** ▶ Solve the equation

$$5 \times n = 60$$

**Solution.** Because 5 multiplies n on the left side, to isolate n, *divide* both sides by 5.

$$\frac{5 \times n}{5} = \frac{60}{5}$$

$$\frac{\overset{1}{\cancel{5} \times n}}{\underset{1}{\cancel{5}}} = \frac{\overset{12}{\cancel{60}}}{\underset{1}{\cancel{5}}}$$

$$n = 12$$

Thus the solution is 12.                                                      ◀

Now here is an equation involving percent.

**Example 3** ▶ Solve the equation

$$50\% \times n = 7$$

**Solution.** Change 50% to .5.

$$.5 \times n = 7$$

To eliminate decimals, multiply both sides by 10.

$$10 \times .5 \times n = 10 \times 7$$

$$5 \times n = 70$$

Divide both sides by 5.

$$\frac{\overset{1}{\cancel{5} \times n}}{\underset{1}{\cancel{5}}} = \frac{\overset{14}{\cancel{70}}}{\underset{1}{\cancel{5}}}$$

$$n = 14$$

Thus the solution is 14.                                                ◀

*Try the exercises for Topic B on page 93.*

## C. 20% OF WHAT NUMBER?

**Example 4** ▶  If 30 is 20% of a certain number, find that number.

*Solution.*  Reword the question as follows.

30 is equal to 20% of *what* number?

Now let *n* stand for the *unknown* number. Recall that "of" often indicates multiplication. Translate the problem as follows:

30 is equal to 20% of what number?

$$30 \quad = \quad 20\% \times \quad n$$

$$30 = .2 \times n \qquad \text{Multiply both sides by 10 to eliminate decimals.}$$

$$10 \times 30 = 10 \times .2 \times n$$

$$300 = 2 \times n \qquad \text{Divide both sides by 2 to isolate the } n.$$

$$\frac{300}{2} = \frac{2 \times n}{2}$$

$$150 = n$$

Thus 30 is 20% of 150.                                                ◀

*Try the exercises for Topic C on page 94.*

## D. WHAT PERCENT?

**Example 5** ▶  What percent of 40 is 30?

(A) 75%      (B) 25%      (C) 30%      (D) 40%      (E) 133.3%

*Solution.*  To see what is involved here, reword the question.

What percent is 30 of 40?

You are really asked here to write the fraction $\frac{30}{40}$ as a decimal. Thus

$$\frac{30}{40} = \frac{3}{4}$$

Divide 3 by 4

$$\begin{array}{r} .75 \\ 4\overline{\smash{)}3.00} \end{array}$$

to obtain

$$\frac{3}{4} = .75 = 75\%$$

Thus 30 is 75% of 40.  The correct choice is (A).    ◄

*Try the exercises for Topic D on page 95.*

## E.  APPLIED PROBLEMS

Problems involving percent arise in many applications. Often you must first reword a problem.

**Example 6 ►**  Sue correctly answered 90% of the 40 problems on her math quiz. How many problems did she answer correctly?

(A) 4          (B) 9          (C) 35          (D) 36          (E) 38

*Solution.*  Reword the problem.

What is 90% of 40?

Here again, the word "of" indicates multiplication. Thus find

$$.9 \times 40 = 36.0 = 36$$

She answered 36 problems correctly. The correct choice is (D).    ◄

**Example 7 ►**  15% of Bob's salary goes into taxes. If his taxes are $30.75 each week, find his weekly salary (before taxes).

*Solution.*  15% of Bob's salary equals taxes. Write in $30.75 for his taxes, and find his salary, $s$.

15% of Bob's salary equals $30.75

15% ×      $s$      =      30.75

$$.15 \times s = 30.75$$

$$100 \times .15 \times s = 100 \times 30.75$$

$$15 \times s = 3075 \qquad \text{Divide both sides by 15.}$$

$$\frac{15s}{15} = \frac{3075}{15}$$

$$s = 205$$

His weekly salary is $205.                                                    ◀

*Try the exercises for Topic E on page 95.*

## EXERCISES

**A.**  *In Exercises 1-6, draw a circle around the correct letter.*

1.  What is 50% of 66?

    (A) 50          (B) 33          (C) 132          (D) 13.2          (E) 99

2.  What is 40% of 8?

    (A) 32          (B) 3.2          (C) 320          (D) 4          (E) 12

3.  What is 75% of 60?

    (A) 15          (B) 42.5          (C) 425          (D) 45          (E) 48

4.  What is 70% of 200?

    (A) 70          (B) 14          (C) 140          (D) 1400          (E) 60

5.  What is 30% of 90?

    (A) 30          (B) 33          (C) 270          (D) 27          (E) 120

6.  What is 80% of 120?

    (A) 100          (B) 96          (C) 90          (D) 108          (E) 24

7.  What is 20% of 1440?   *Answer:*        8.  What is 60% of 55?   *Answer:*

9.  What is 35% of 80?   *Answer:*        10.  What is 65% of 90?   *Answer:*

11.  What is 85% of 80?   *Answer:*        12.  What is 17% of 60?   *Answer:*

**B.**  *Solve each equation for n.*

13.  $2n = 20$          *Answer:*          14.  $3n = 36$          *Answer:*

15.  $4n = 28$          *Answer:*          16.  $10n = 130$          *Answer:*

17.  $7n = 56$          *Answer:*          18.  $5n = 75$          *Answer:*

19.  $20n = 140$          *Answer:*          20.  $8n = 60$          *Answer:*

21.  $50\% \times n = 9$          *Answer:*          22.  $40\% \times n = 6$          *Answer:*

23.  $30\% \times n = 27$        *Answer:*        24.  $80\% \times n = 120$        *Answer:*

25.  $75\% \times n = 15$        *Answer:*        26.  $60\% \times n = 30$        *Answer:*

27.  $90\% \times n = 270$        *Answer:*        28.  $35\% \times n = 105$        *Answer*

C.    *In Exercises 29-34, draw a circle around the correct letter.*

29.  20% of what number is 12?

    (A) 240        (B) 24        (C) 204        (D) 60        (E) 14.4

30.  25% of what number is 8?

    (A) 2        (B) 10        (C) 200        (D) 32        (E) 33

31.  60% of what number is 24?

    (A) 14.4        (B) 144        (C) 40        (D) 84        (E) 15

32.  75% of what number is 84?

    (A) 112        (B) 125        (C) 21        (D) 63        (E) 105

33.  If 36 is 25% of a certain number, find that number.

    (A) 9        (B) 27        (C) 61        (D) 48        (E) 144

34.  If 40% of a certain number is 20, find that number.

    (A) 8        (B) 80        (C) 800        (D) 50        (E) 60

35.  If 90% of a certain number is 90, find that number.  *Answer:*

36.  If 30% of a certain number is 18, find that number.  *Answer:*

37.  If 25% of a number is 120, what is the number?  *Answer:*

38.  If 75% of a number is 72, what is the number?  *Answer:*

39.  If 45% of a number is 4.5, what is the number?  *Answer:*

40.  If 22% of a number is 33, what is the number?  *Answer:*

D.    *In Exercises 41-44, draw a circle around the correct letter.*

41. What percent of 50 is 25?

    (A) 50%        (B) 25%        (C) 2%        (D) 20%        (E) 200%

42. What percent of 48 is 36?

    (A) $33\frac{1}{3}$%      (B) 133%        (C) 25%        (D) 75%        (E) 125%

43. What percent of 90 is 72?

    (A) 72%        (B) 90%        (C) 20%        (D) 80%        (E) 120%

44. What percent of 40 is 12?

    (A) 30%        (B) 20%        (C) 12%        (D) 48%        (E) 52%

45. What percent of 80 is 72?      *Answer:*

46. What percent of 125 is 75?      *Answer:*

47. What percent of 36 is 3.6?      *Answer:*

48. What percent of 24 is 36?      *Answer:*

E.  *In Exercises 49-54, draw a circle around the correct letter.*

   49. A student answers 82% of the questions on an exam correctly. If there are 50
       questions on the exam, how many does she answer correctly?

       (A) 50        (B) 40        (C) 41        (D) 42        (E) 32

   50. Harry answers 36 out of 60 questions correctly. What percent of his answers are
       correct?

       (A) 36%        (B) 50%        (C) 54%        (D) 60%        (E) 72%

   51. A pitcher throws 50 called strikes out of a total of 125 pitches in a game. What
       percent of his pitches are called strikes?

       (A) 25%        (B) 40%        (C) 50%        (D) 85%        (E) 250%

   52. An alloy contains 60% copper. How much copper is there in 400 tons of the alloy?

       (A) 24 tons    (B) 240 tons    (C) 640 tons    (D) 320 tons    (E) $666\frac{2}{3}$ tons

   53. 20% of the people in a village are retired. If there are 120 retired people there, what
       is the population of the village?

       (A) 24        (B) 2400        (C) 24,000        (D) 600        (E) 144

54. A family pays $550 a month for rent. This amounts to 25% of their monthly income. What is their monthly income?

(A) $112.50    (B) $662.50    (C) $2000    (D) $2200    (E) $1375

55. A basketball player made 80% of her foul shot attempts. If she made 16 foul shots, how many did she attempt?    *Answer:*

56. Al pays 14% of his weekly salary in taxes. If his taxes are $22.40, his weekly salary (before taxes) is _____

57. An airplane pilot earns $38,000 per year. 22% of his salary is withheld for taxes. How much money is withheld?    *Answer:*

58. A man pays a sales tax of $13.05 on a suit. The sales tax rate is 9%. What is the price of the suit (before the tax)?    *Answer:*

59. At a certain college, 35% of the students are enrolled in at least one mathematics course. If 2310 students are enrolled in at least one math course, how many students are there at the college?    *Answer:*

60. If the sales tax on a $4.50 item is $.36, what is the sales tax rate?    *Answer:*

# 15.  Percent Increase or Decrease

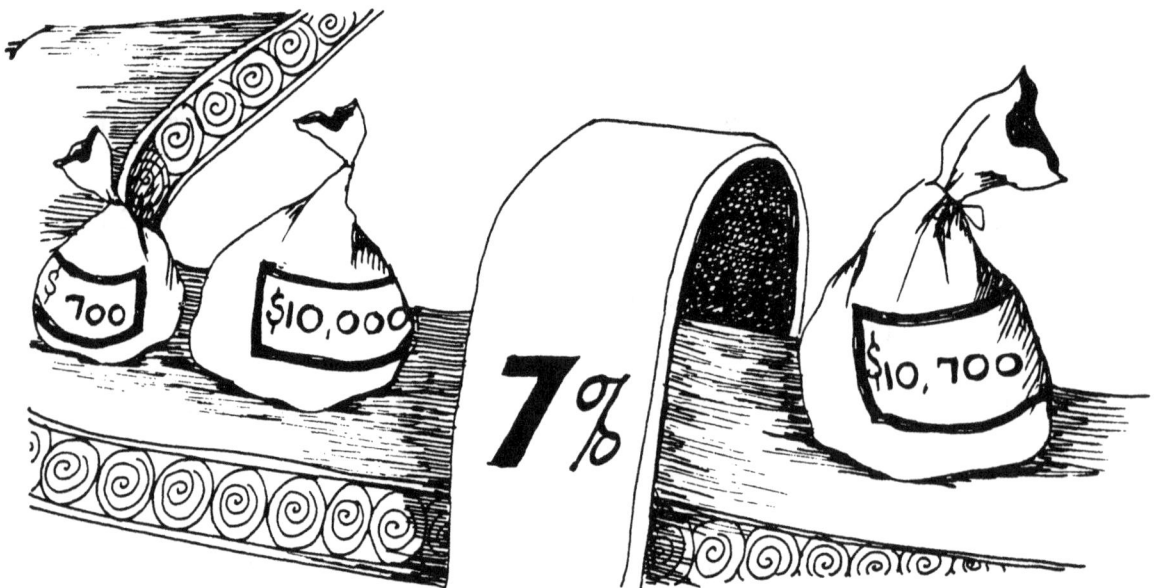

Percent increase or decrease arises in applications such as sales tax, price fluctuation, and population shifts.

## A. PERCENT INCREASE AND SALES TAX

A part-time student earns $10,000 a year. He receives a 7% *increase* in salary. The  increase amounts to

$$\$10,000 \times .07 = \$700$$

This *increase* is in *addition* to *all* of (100% of) his former salary. In other words, he receives 100% of his former salary *plus* an additional 7%. Thus his new salary is 107% of his former salary. To determine his new salary, multiply $10,000 by 107%.

$$\$10,000 \times 107\% = \$10,000 \times 1.07$$
$$= \$10,700$$

His new salary is $10,700.

Similarly, for an increase of 8%, multiply the former salary by 108% or 1.08. For an increase of 10%, multiply by 110% or 1.1.

**Example 1 ▶** A jacket sells for $80 plus a 6% sales tax. What is the total price?

(A) $86        (B) $4.80        (C) $75.20        (D) $84.80        (E) $140

*Solution.* The 6% sales tax is *in addition to* the price of the jacket. Thus 100% of the original price is increased by 6%. To find the *total* price, multiply the price of the jacket, $80, by 1.06.

```
    1.06  ◄──── 2 decimal digits
 X  $80   ◄──── + 0 decimal digits
  $84.80  ◄──── 2 decimal digits
```

The total price is $84.80, choice (D).                                    ◄

**Example 2 ▶** The price of coffee, which was originally $2.80 per pound, is increased by 5%. What is the new price?

*Solution.* An *increase* of 5% means you must multiply the original price by 105% or 1.05.

```
    $2.80  ◄──── 2 decimal digits
 X   1.05  ◄──── + 2 decimal digits
    14 00
   2 80 0
  $2.94 00 ◄──── 4 decimal digits
```

Because 2.9400 = 2.94, the new price is $2.94.                            ◄

*Try the exercises for Topic A on page 98.*

## B. PERCENT DECREASE AND REDUCTIONS

During hard times, a man who was earning $10,000 a year is asked to take a 5% *reduction* in his salary. The *reduction* (or *decrease*) amounts to

$10,000 × .05, or $500

This decrease is *subtracted from* 100% of his former salary. Thus he receives 100% of his former salary *minus* the 5% reduction. His new salary is therefore only 95% of his former salary. To determine his new salary multiply $10,000 by 95%.

$$\$10,000 \times 95\% = \$10,000 \times .95$$
$$= \$9500$$

His new salary is $9500.

**Example 3 ▶** A town with 24,000 people loses 6% of its population. What is the new population?

*Solution.*

100% - 6% = 94% = .94

```
   240 00  ———— 0 decimal digits
  X  .94  ——— + 2 decimal digits
   960 00
  21600 0
  22560.00  ———— 2 decimal digits
```

The new population is 22,560.    ◀

**Example 4 ▶** At a sale, a $28 sweater is reduced by 20%. Its sale price is

(A) $48      (B) $8      (C) $33.60      (D) $23.40      (E) $22.40

*Solution.*

100% - 20% = 80% = .80

$28 × .80 = $22.40

The sale price is $22.40, choice (E).    ◀

*Try the exercises for Topic B on page 99.*

## EXERCISES

A.   *In Exercises 1-6, draw a circle around the correct letter.*

1.   Bob's weekly paycheck, which used to be $200, is increased by 12%. His new paycheck is

(A) $212      (B) $224      (C) $2400      (D) $188      (E) $200.12

2.  Jane's monthly paycheck, which used to be $1800, is increased by 8%. Her new paycheck is

(A) $1944      (B) $1656      (C) $144      (D) $1808      (E) $1814.44

3.  A coat sells for $140 plus a 7% sales tax. What is the total price?

(A) $140.98      (B) $1498      (C) $149.80      (D) $147      (E) $980

4.  A tie sells for $7.50 plus a 6% sales tax. What is the total price?

(A) $7.56      (B) $8.10      (C) $7.95      (D) $8.00      (E) $7.92

5.  The price of gasoline, which was $1.25 per gallon, is increased by 4%. The new price per gallon is

(A) $1.29      (B) $6      (C) $1.21      (D) $1.30      (E) $1.20

6.  A business that makes a profit of $180,000 one year increases its profit by 25% the next year. Its profit is then

(A) $45,000      (B) $25,000      (C) $205,000    (D) $225,000    (E) $135,000

7.  A man who earns $46,000 per year receives a 4% increase in salary. What is his new salary?    *Answer:*

8.  The population of a city increases by 5%. If the population was formerly 4,400,000, its new population is ————————————

9.  **A store increases the price of its chocolates by 12%. If the old price was $4.50 per pound, the new price per pound is ————————————**

10.  Season's ticket sales for Laker games increased by 10% this season. If last season the Lakers sold 4230 season's tickets, how many do they sell this season?

*Answer:*

11.  **A saleswoman who earns $26,600 per year receives a 6% increase in salary. Her new salary is ————————————**

12.  **The price of milk, which was $2.20 per gallon, is increased by 5%. The new price is**

————————————

B.  *In Exercises 13-18, draw a circle around the correct letter.*

13.  **A man who was earning $20,000 per year receives a 2% reduction in his salary. His new salary is**

(A) $18,000      (B) $19,800      (C) $16,000      (D) $19,840      (E) $19,600

14.  **A town with 10,000 people loses 8% of its population. Its new population is**

(A) 9200      (B) 2000      (C) 8000      (D) 9920      (E) 9992

15.  **A sweater that sold for $20 was reduced by 15%. Its new price is**

(A) $18.50      (B) $5.00      (C) $15.00      (D) $17.00      (E) $17.50

16. **A coat that sold for $199 was reduced by 10%. Its new price is**

(A) $180.10    (B) $179.10    (C) $19.90    (D) $189    (E) $179

17. **An alarm clock that sold for $29.60 was reduced by 25%. Its new price is**

(A) $22.10    (B) 27.10    (C) $22.20    (D) $22.00    (E) $27.00

18. **A business that makes a profit of $80,000 one year decreased its profit by 30% the next year. Its new profit is**

(A) $50,000    (B) $56,000    (C) $24,000    (D) $104,000    (E) $79,760

19. **A town with 180,000 people loses 3% of its population. What is its new population?**

*Answer:*

20. **A mine produces 450 million tons of ore one year. The next year there is a 6% decrease in production. How many tons of ore are produced the next year?**

*Answer:*

21. **A worker who earns $16,000 has 12% deducted from his salary for taxes and union dues. What is his take-home pay?**

*Answer:*

22. **A suit that sold for $97.50 was reduced by 8%. Its new price is**

*Answer:*

# 16.  Area and Perimeter

Arithmetic methods are used to solve problems concerning the area or the perimeter of a rectangular region. Sometimes these problems also involve a cost factor (*Topics C and E*).

## A. SQUARE OF A NUMBER

DEFINITION

> The **square** of a number means the number times itself.

Write

$$5^2 \qquad \text{for} \qquad \text{the square of 5}$$

Thus

$$5^2 = 5 \times 5 = 25$$

Similarly,

$$10^2 = 10 \times 10 = 100$$

**Example 1** ▶ Find:

(a) $8^2$       (b) $12^2$

*Solution*.

(a) $8^2 = 8 \times 8 = 64$         (b) $12^2 = 12 \times 12 = 144$      ◀

*Try the exercises for Topic A on page 107.*

## B. AREA OF A RECTANGLE

Area involves two "dimensions" and is measured in terms of *square* inches, *square* feet, and so on. We will write

in.$^2$     for     in. $\times$ in.     [square inches]
ft.$^2$     for     ft. $\times$ ft.     [square feet]

The area of a rectangle equals its length times its width.

> Area of Rectangle = length $\times$ width

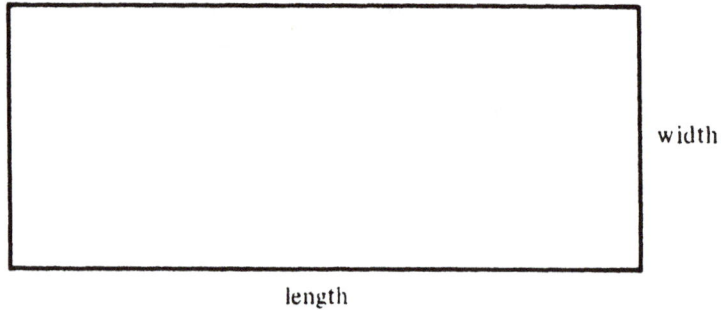

width

length

For example, if the length is 5 inches and the width is 2 inches, then

Area = length $\times$ width

= 5 in. $\times$ 2 in.

= 10 in.$^2$

2 inches

5 inches

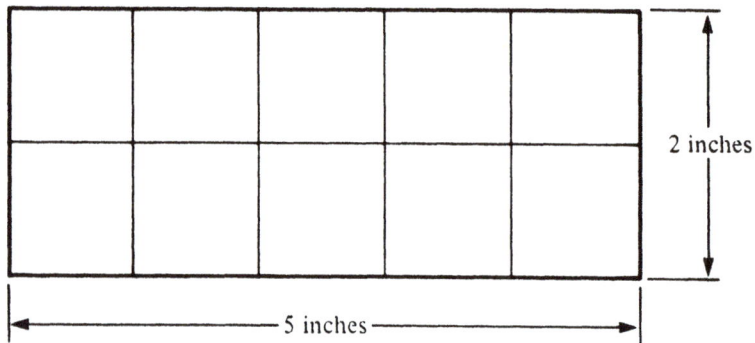

In the above figure, each box represents one square inch. Altogether, there are 10 square inches. Thus

5 in. $\times$ 2 in. = 10 in.$^2$

**Example 2** ▶ Find the area of a rectangular parking lot that is 100 feet long and 60 feet wide.

(A) 160 ft.     (B) 320 ft.     (C) 600 ft.$^2$     (D) 6000 ft.$^2$     (E) 13,600 ft.$^2$

***Solution.***

$$\text{Area} = \text{length} \times \text{width}$$
$$= 100 \text{ ft.} \times 60 \text{ ft.}$$
$$= 6000 \text{ ft.}^2 \qquad (square \text{ feet})$$

The area is 6000 ft.². The correct choice is (D).          ◄

In the case of a vertical rectangular region, such as a wall or a television screen, the area is often given in terms of *height* and width.

Area = height × width

**Example 3** ▶ Find the area of a rectangular movie screen that is 20 feet high and 30 feet wide.

(A) 600 ft.²   (B) 6000 ft.²   (C) 50 ft.   (D) 100 ft.   (E) 1300 ft.²

***Solution.***

$$\text{Area} = \text{height} \times \text{width}$$
$$= 20 \text{ ft.} \times 30 \text{ ft.}$$
$$= 600 \text{ ft.}^2$$

Choice (A) is correct.          ◄

Some problems concern *solid* (3-dimensional) rectangular figures, such as boxes, and you must select the two dimensions for the area in question.

**Example 4** ▶ A rectangular dresser is 28 inches long, 20 inches wide, and 40 inches high. Find the area of its base.

***Solution.*** The base is the *bottom* of the dresser. The height is of no concern in this problem. Thus

$$\text{Area} = \text{length} \times \text{width}$$
$$= 28 \text{ in.} \times 20 \text{ in.}$$
$$= 560 \text{ in.}^2$$

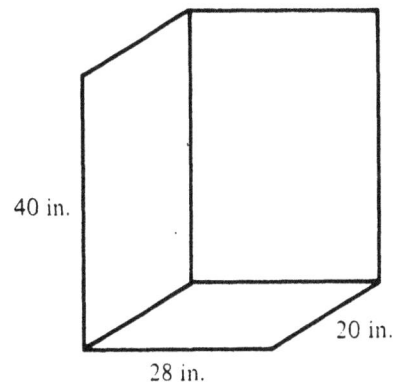

40 in.          20 in.

28 in.          ◄

Sometimes you are given the area of a rectangle as well as *either* the length or width. You are then asked to find the other dimension.

**Example 5** ▶ The area of a rectangular tabletop is 35 square feet. If the length is 7 feet, find the width.

*Solution.*

$$\text{Area} = \text{length} \times \text{width}$$

$$35 \text{ ft.}^2 = 7 \text{ ft.} \times \text{width}$$

Divide both sides by 7 ft. and note that

$$\frac{\text{ft.}^2}{\text{ft.}} = \frac{\overset{1}{\cancel{\text{ft.}}} \times \text{ft.}}{\underset{1}{\cancel{\text{ft.}}}} = \text{ft.}$$

$$\frac{35 \text{ ft.}^2}{7 \text{ ft.}} = \text{width}$$

$$5 \text{ ft.} = \text{width} \qquad\qquad ◄$$

*Try the exercises for Topic B on page 107.*

## C. AREA AND COST

**Example 6** ▶ What is the cost of the material for a tablecloth that is to be 8 feet by 6 feet, if the material costs $4 per square foot?

*Solution.* First find the area of the tablecloth.

$$\text{Area} = \text{length} \times \text{width}$$

$$= 8 \text{ ft.} \times 6 \text{ ft.}$$

$$= 48 \text{ ft.}^2$$

To find the cost, now multiply 48 square feet by the cost of $4 per square foot. Note that $4 *per* square foot can be expressed by the fraction $\frac{\$4}{\text{ft.}^2}$.

$$\text{Cost} = 48 \overset{1}{\cancel{\text{ft.}^2}} \times \frac{\$4}{\underset{1}{\cancel{\text{ft.}^2}}} = \$192 \qquad\qquad ◄$$

*Try the exercises for Topic C on page 108.*

## D. PERIMETER

DEFINITION

> The **perimeter** of a closed figure, such as a rectangle, is the sum of the lengths of its sides.

In other words, the perimeter of a closed figure is the length of its boundary.

$$\text{Perimeter of Rectangle} = \text{Length} + \text{Width} + \text{Length} + \text{Width}$$
$$= 2 \times \text{Length} + 2 \times \text{Width}$$

2 is a common factor on the right side. Thus

$$\boxed{\text{Perimeter of Rectangle} = 2(\text{Length} + \text{Width})}$$

In the accompanying figure, the length is 12 inches and the width is 4 inches.

$$\text{Perimeter} = 2(\text{length} + \text{width})$$
$$= 2(12 \text{ in.} + 4 \text{ in.})$$
$$= 2(16 \text{ in.})$$
$$= 32 \text{ in.}$$

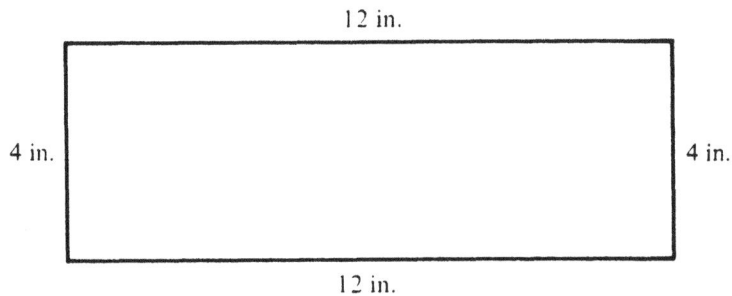

12 in.

4 in.                4 in.

12 in.

Note that perimeter measures boundary length (and *not* area). Units of perimeter are inches, feet, and yards (and *not* square inches, and so on).

**Example 7** ▶ Find the perimeter of a rectangle that is 9 feet long and 8 feet wide.

(A) 17 ft.    (B) 72 ft.    (C) 34 ft.    (D) 72 ft.²    (E) 144 ft.

*Solution.*

$$\text{Perimeter} = 2(\text{length} + \text{width})$$
$$= 2(9 \text{ feet} + 8 \text{ feet})$$
$$= 2(17 \text{ feet})$$
$$= 34 \text{ feet}$$

The correct choice is (C).    ◀

**Example 8** ▶ The perimeter of a rectangle is 30 yards and the width is 5 yards. Find the length.

*Solution.*

$$\text{Perimeter} = 2(\text{length} + \text{width})$$
$$30 \text{ yd.} = 2(\text{length} + 5 \text{ yd.})$$

Divide both sides by 2.

$$\frac{\overset{15}{\cancel{30}} \text{ yd.}}{\underset{1}{\cancel{2}}} = \frac{\overset{1}{\cancel{2}}(\text{length} + 5 \text{ yd.})}{\underset{1}{\cancel{2}}}$$

$$15 \text{ yd.} = \text{length} + 5 \text{ yd.}$$

Subtract 5 yd. from each side.

$$15 \text{ yd.} - 5 \text{ yd.} = \text{length} + 5 \text{ yd.} - 5 \text{ yd.}$$
$$10 \text{ yd.} = \text{length}$$

◀

*Try the exercises for Topic D on page 109.*

## E. PERIMETER AND COST

**Example 9** ▶ A woman wants to fence in a garden that is 20 feet by 15 feet. If fencing costs $6 per foot, how much will the fencing cost?

*Solution.* Here perimeter (and *not* area) is involved because the fence goes around the *boundary* of the garden. First find the perimeter.

$$\text{Perimeter} = 2(\text{length} + \text{width})$$
$$= 2(20 \text{ ft.} + 15 \text{ ft.})$$
$$= 2(35 \text{ ft.})$$
$$= 70 \text{ ft.}$$

Now find the cost. Multiply 70 feet by $6 *per* foot. Note that $6 per foot can be expressed by the fraction

$$\frac{\$6}{\text{ft.}}$$

$$\text{Cost} = 70 \overset{1}{\cancel{\text{ft.}}} \times \frac{\$6}{\underset{1}{\cancel{\text{ft.}}}}$$

$$= \$420$$

◀

*Try the exercises for Topic E on page 110.*

**EXERCISES**

**A.** *Find each square.*

1.  $4^2$          *Answer:*          2.  $6^2$          *Answer:*

3.  $7^2$          *Answer:*          4.  $9^2$          *Answer:*

5.  $12^2$         *Answer:*          6.  $0^2$          *Answer:*

**B.**    *When several choices are given, draw a circle around the correct letter.*

7.  Find the area of a rug that is 6 feet by 3 feet.

   (A) 9 ft.²      (B) 12 ft.²      (C) 18 ft.²      (D) 18 ft.      (E) 36 ft.²

8.  Find the area of a tabletop that is 40 inches long and 20 inches wide.

   (A) 80 in.²      (B) 60 in.²      (C) 120 in.²      (D) 800 in.²      (E) 8000 in.²

9.  Find the area of a blotter that is 35 inches long and 20 inches wide.

   (A) 55 in.      (B) 110 in.      (C) 70 in.²      (D) 700 in.²      (E) 7000 in.²

10.  Find the area of a floor that is 20 feet by 14 feet.

   (A) 280 ft.²      (B) 560 ft.²      (C) 68 ft.²      (D) 340 ft.²      (E) 680 ft.²

11.  Find the area of a page that is 9 inches by 6 inches.  *Answer:*

12.  Find the area of a wall that is 9 feet high and 12 feet wide.  *Answer:*

13.  Find the area of a rectangular television screen that is 11 inches high and 20 inches wide.  *Answer:*

14.  Find the area of a rectangular mirror that is 7 feet high and 2 feet wide.
    *Answer:*

15.  Find the area of the base of a rectangular cabinet that is 30 inches long, 10 inches wide, and 14 inches high.  *Answer:*

16.  Find the area of the floor of a room that is 16 feet long, 12 feet wide, and 10 feet high.  *Answer:*

17.  Find the area of one wall of a room that is 15 feet long, 15 feet wide, and 9 feet high.  *Answer:*

18.  Find the area of the ceiling of a room that is 22 feet long, 20 feet wide, and 10 feet high.  *Answer:*

19. Find the length of a rectangle whose width is 9 yards and whose area is 90 square yards.

    (A) 10 yd.      (B) 10 yd.²      (C) 81 yd.      (D) 36 yd.      (E) 810 yd.²

20. The length of a rectangle is 11 feet and the area is 88 square feet. Find the width.

    (A) 8 ft.      (B) 77 ft.      (C) 38.5 ft.      (D) 99 ft.      (E) 968 ft.

21. The area of a rectangular carpet is 24 square feet and the length is 8 feet. Find the width. *Answer:*

22. The area of a rectangular bedspread is 54 square feet and the width is 6 feet. Find the length. *Answer:*

C.  *In Exercises 23-26, draw a circle around the correct letter.*

23. What does it cost to carpet a room that is 15 feet by 10 feet, if the carpeting costs $3 per square foot?

    (A) $150      (B) $45      (C) $30      (D) $450      (E) $75

24. What does it cost to carpet a room that is 7 yards by 5 yards, if the carpeting costs $12 per square yard?

    (A) $144      (B) $420      (C) $35      (D) $300      (E) $600

25. Find the cost of a tablecloth that is 5 yards by 3 yards at 12 dollars per square yard.

    (A) $150      (B) $96      (C) $192      (D) $180      (E) $360

26. It costs 20 cents per square yard to varnish a floor. How much will it cost to varnish the floor of a hall that is 40 yards by 10 yards?

    (A) $40      (B) $400      (C) $800      (D) $80      (E) $8000

27. A blanket is to be 7 feet by 4 feet. The material for the blanket costs 3 dollars per square foot. Find the cost of the material for this blanket. *Answer:*

28. Glass costs $2.50 per square yard. How much will a pane of glass that is 6 feet long and 3 feet wide cost? *Answer:*

29. A woman buys a piece of material that is 20 feet by 3 feet. If the material costs $3.50 per square foot, how much does she pay? *Answer:*

30. Carpeting costs $12.50 per square foot. How much will it cost to carpet a floor that is 15 feet by 12 feet? *Answer:*

31.  It costs $2700 to carpet a floor that is 20 feet by 15 feet. How much does this carpeting cost per square foot? *Answer:*

32.  Material for a tablecloth that is to be 6 feet by 4 feet costs $132. How much does the material cost per square foot? *Answer:*

D.  *When several choices are given, draw a circle around the correct letter.*

33.  Find the perimeter of a rectangle whose length is 8 inches and whose width is 5 inches.

   (A) 13 in.       (B) 26 in.       (C) 40 in.       (D) 20 in.       (E) 80 in.

34.  Find the perimeter of a rectangle that is 3 yards by 2 yards.

   (A) 5 yd.       (B) 6 yd.       (C) 10 yd.       (D) 12 yd.       (E) 24 yd.

35.  A woman wants to sew a silk border around a blanket that is 7 feet by 4 feet. How long a piece of silk does she need?

   (A) 11 ft.       (B) 15 ft.       (C) 22 ft.       (D) 28 ft.       (E) 56 ft.

36.  How much fencing is needed to fence in a park that is 75 feet long and 35 feet wide?

   (A) 110 ft.       (B) 220 ft.       (C) 2625 ft.       (D) 2625 ft.²   (E) 262.5 ft.

37.  How long a rope is needed to enclose a rectangular parking lot that is 100 yards by 54 yards? *Answer:*

38.  A rectangular field is 400 yards by 300 yards. To enclose the field would take how many yards of fencing? *Answer:*

39.  A rectangle has length 5 inches and perimeter 18 inches. Find its width.

   (A) 13 in.       (B) 23 in.       (C) $\frac{18}{5}$ in.       (D) 6.5 in.       (E) 4 in.

40.  A rectangle has width 10 feet and perimeter 50 feet. Find its length.

   (A) 5 ft.       (B) 40 ft.       (C) 15 ft.       (D) 10 ft.       (E) 25 ft.

41.  A rectangle is 14 inches long. If its perimeter is 40 inches, what is its width?

   *Answer:*

42.  The perimeter of a rectangle is 100 inches. If its width is 10 inches, its length is

_____

43.  The perimeter of a rectangle is 20 inches and its length is 6 inches. Its *area* is

_____

**44.** The area of a rectangle is 20 square inches and its length is 5 inches. Find its *perimeter.* *Answer:*

**E.** *In Exercises 45-48, draw a circle around the correct letter.*

**45.** Fencing costs $5 per foot. How much does it cost to fence in a rectangular garden that is 20 feet by 10 feet?

   (A) $100      (B) $500      (C) $300      (D) $1000      (E) $2000

**46.** A ribbon is used as the border of a blanket that is 6 feet by 4 feet. If the ribbon costs $2 per foot, what is the cost of the border?

   (A) $24      (B) $48      (C) $12      (D) $20      (E) $40

**47.** A rectangular field is enclosed by wiring that costs $.50 per yard. If the field is 900 yards by 500 yards, the cost of the wiring is

   (A) $450,000   (B) $225,000   (C) $700      (D) $1400      (E) $2800

**48.** A fence is needed for a rectangular park that is 50 yards by 40 yards. If the fencing costs $4 per yard, the total cost of the fencing is

   (A) $8000      (B) $800      (C) $4000      (D) $360      (E) $720

**49.** A garden that is 20 feet by 12 feet is fenced in for $1600. What is the cost per foot of the fencing? *Answer:*

**50.** A rectangular lot is fenced in for $1000. If the fencing costs $5 per foot and the lot is 40 feet wide, how long is the yard? *Answer:*

**51.** Piping that goes around a rectangular rooftop 40 feet long and 30 feet wide costs $3 per foot. What is the cost of the piping? *Answer:*

**52.** A rope encloses a football field that is 100 yards long and 53 yards wide. If the rope costs $1.50 per yard, what is the total cost? *Answer:*

# 17.  Graphs

It is important to know how to interpret a graph that might appear in a newspaper, magazine, or textbook.

## A. BAR GRAPHS

In a bar graph the *height* of a column or bar represents a number.

**Example 1 ▶**  Use the graph at the right to find out the total number of magazines sold by Collegiate Press from April through June, 1983.

(A) 3000      (B) 3500

(C) 3.5       (D) 9.5

(E) 9500

1983 Sales
(Thousands)

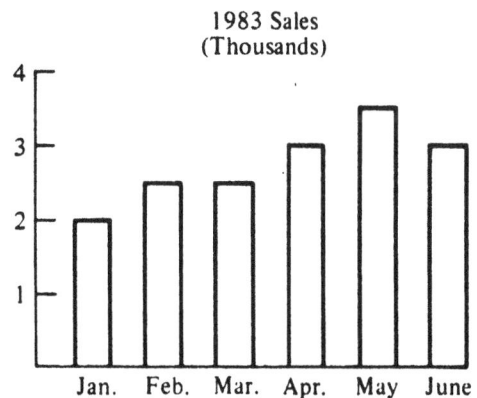

*Solution.* Each bar represents the number of *thousands* of magazines sold each month from January through June, 1983. To find how many thousands were sold, compare the top of the bar with the scale at the left. In April, 3000 copies were sold and in June, 3000 copies were sold. In May, 3500 copies were sold. (The top of the bar for May lies midway between the 3 and the 4 at the left. Thus the bar represents 3500, which is midway between 3000 and 4000. Note that 3500 is the average of 3000 and 4000.) Altogether there were 9500 copies sold during the months April, May, and June, 1983. The correct choice is (E).  ◀

*Try the exercises for Topic A on page 113.*

## B. CIRCLE GRAPHS

In a circle graph, various items are presented as a *percentage* of the total dollar amount.

**Example 2** ▶ In the graph at the right, find the annual expenditures for schools.

(A) 10%    (B) $15,500,000

(C) $1,550,000

(D) $155,000    (E) $155,000,000

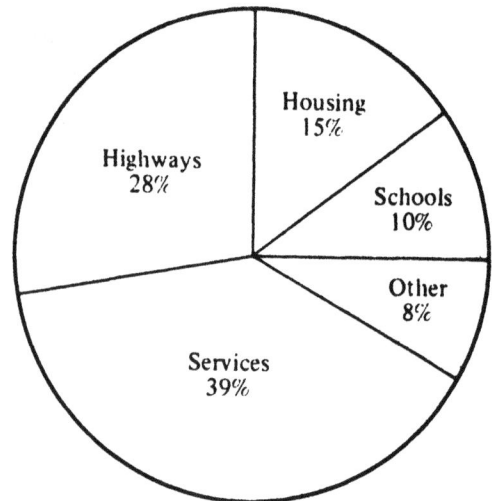

Annual Expenditures: $15,500,000

*Solution.* The graph indicates that 10% of the annual expenditure of $15,500,000 is spent on schools. Because 10% = .1,

$$10\% \times \$15,500,000 = .1 \times \$15,500,000$$

$$= \$1,550,000$$

Choice (C) is correct.  ◀

*Try the exercises for Topic B on page 116.*

## C. LINE GRAPHS

Line graphs are particularly useful when you want to compare two successive periods.

**Example 3** ▶ The graph to the right shows the number of students enrolled in colleges in New England from 1975 to 1979. What was the increase in enrollment from 1977 to 1978?

Enrollment in
New England Colleges

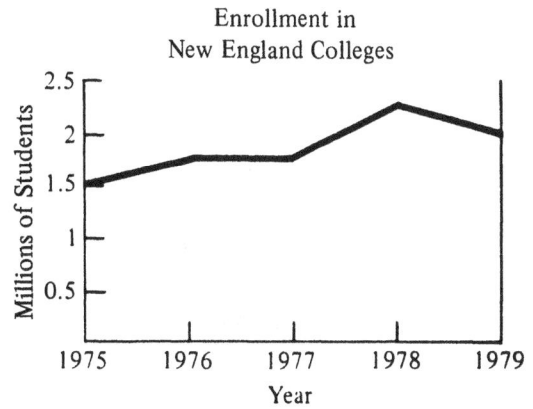

(A) 250,000   (B) 500,000   (C) 750,000   (D) 1,750,000   (E) 2,250,000

*Solution.* In 1977 there were 1,750,000 students enrolled. In 1978 there were 2,250,000. To find the *increase* in enrollment, subtract:

$$\begin{array}{r} 2\ 250\ 000 \\ -\ 1\ 750\ 000 \\ \hline 500\ 000 \end{array}$$

The increase in enrollment was 500,000 [Choice (B)].                    ◀

*Try the exercises for Topic C on page 117.*

## EXERCISES

**A.** *Draw a circle around the correct letter.*

*For Exercises 1-5 use the graph at the right, which shows the number of students who graduated from Truman College each year from 1976 to 1980.*

**GRADUATES**

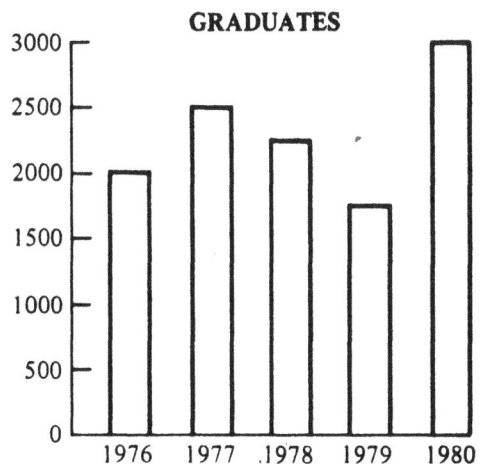

1. The number that graduated in 1977 was

(A) 2000                (B) 2500                (C) 2250

(D) 12,500,000          (E) 250,000

2. The number that graduated in 1979 was closest to

(A) 1500        (B) 1750        (C) 2000        (D) 2250        (E) 8500

3.  The *total* number that graduated in the two years 1976 and 1977 was closest to

(A) 2000      (B) 2500      (C) 4000      (D) 4500      (E) 5000

4.  The *total* number that graduated in the two years 1978 and 1979 was closest to

(A) 3000      (B) 3500      (C) 4000      (D) 4500      (E) 5000

5.  The year in which the *smallest* number of students graduated was

(A) 1976      (B) 1977      (C) 1978      (D) 1979      (E) 1980

*For Exercises 6–10 use the graph at the right, which shows the number of students attending colleges in New York City in 1980, by borough.*

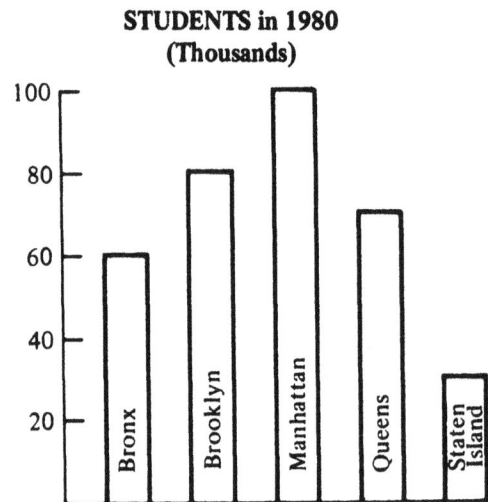

STUDENTS in 1980
(Thousands)

6.  The number attending colleges in Queens in 1980 was closest to

(A) 70           (B) 80

(C) 60,000      (D) 70,000      (E) 80,000

7.  The *total* number attending colleges in Bronx and Brooklyn in 1980 was closest to

(A) 60,000      (B) 80,000      (C) 70,000      (D) 140,000      (E) 100,000

8.  The *total* number attending colleges in Manhattan and Queens was closest to

(A) 70,000      (B) 80,000      (C) 100,000      (D) 170,000      (E) 180,000

9.  How many more students attended a college in Manhattan than in Brooklyn?

(A) 100,000      (B) 80,000      (C) 180,000      (D) 10,000      (E) 20,000

10. The *total* number attending colleges in all five boroughs in 1980 was closest to

(A) 340,000      (B) 320,000      (C) 300,000      (D) 35,000      (E) 40,000

*For Exercises 11–15 use the graph at the right, which shows the monthly income of the XYZ Corporation in 1982 in millions of dollars.*

**1982 INCOME
(Millions)**

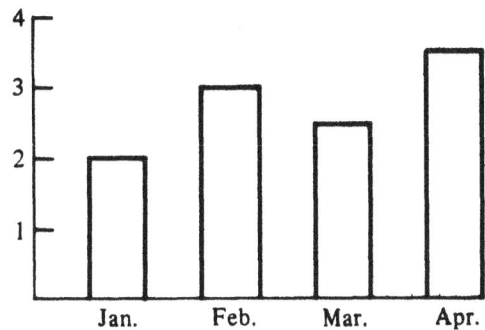

11. The *total* income for January and February was

    (A) $2,000,000    (B) $3,000,000

    (C) $4,000,000    (D) $5,000,000

    (E) $5000

12. The total income for February and March was closest to

    (A) 5,000,000          (B) $5,500,000          (C) $6,000,000

    (D) $6,500,000          (E) $7,000,000

13. The total income for February, March, and April was closest to

    (A) $8,000,000          (B) $8,500,000          (C) $9,000,000

    (D) $9,500,000          (E) $10,000,000

14. How much more income was there in February than in January?

    (A) $500,000          (B) $1,000,000          (C) $1,500,000

    (D) $2,000,000          (E) $3,000,000

15. How much more income was there in April than in March?

    (A) $250,000          (B) $500,000          (C) $750,000

    (D) $1,000,000          (E) $6,000,000

**B.** *Draw a circle around the correct letter.*

*For Exercises 16–20 use the graph at the right, which indicates a family's allotment of their monthly income of $2500.*

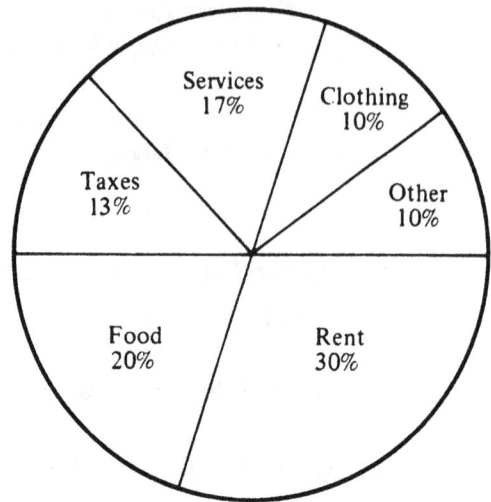

16. How much does the family spend each month for rent?

    (A) $750        (B) $75

    (C) $7500       (D) $1250       (E) $775

17. How much money goes for taxes each month?

    (A) $300     (B) $325     (C) $350     (D) $400     (E) $500

18. Each month the family spends $500 for

    (A) clothing    (B) services    (C) taxes    (D) food    (E) rent

19. What is the *total* amount the family spends each month for food and rent?

    (A) $500     (B) $750     (C) $1000     (D) $1250     (E) $1500

20. How much does the family spend a *year* for clothing?

    (A) $250     (B) $2500     (C) $750     (D) $300     (E) $3000

*For Exercises 21–25 use the graph at the right, which indicates the after-tax disbursements of a company during one year.*

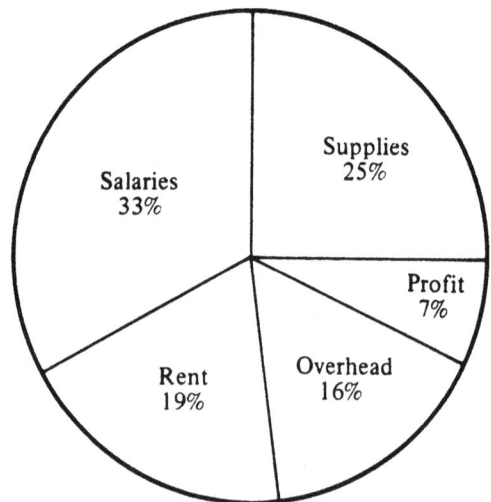

DISBURSEMENTS
$3,600,000

21. How much did the company spend for supplies?

    (A) $250,000    (B) $800,000

    (C) $900,000    (D) $2,500,000

    (E) $9,000,000

22. What was the company's profit that year?

    (A) $252,000          (B) $25,200          (C) $2,520,000

    (D) $70,000           (E) $700,000

23. The total spent for rent and supplies was

(A) $684,000           (B) $1,188,000           (C) $1,872,000

(D) $1,584,000         (E) $2,000,000

24. For which item did the company spend $576,000?

(A) supplies           (B) salaries             (C) rent

(D) overhead           (E) profit

25. How much more did the company spend for salaries than for supplies?

(A) $900,000           (B) $1,188,000           (C) $288,000

(D) $252,000           (E) $324,000

C. *Draw a circle around the correct letter.*

*For Exercises 26–30 use the graph at the right, which indicates the annual profit of the Honest Lenders Finance Corporation, in thousands of dollars.*

CORPORATE PROFITS

26. What was the profit in 1981?

(A) $200,000      (B) $250,000

(C) $250          (D) $400,000

(E) $550,000

27. What was the increase in profit from 1978 to 1979?

(A) $300,000   (B) $400,000   (C) $700,000   (D) $350,000   (E) $100,000

28. What was the increase in profit from 1981 to 1982?

(A) $50,000    (B) $100,000   (C) $150,000   (D) $300,000   (E) $350,000

29. What was the *decrease* in profit from 1979 to 1980?

(A) $150,000   (B) $200,000   (C) $250,000   (D) $400,000   (E) **$600,000**

**30.** What was the total profit for the years 1979 and 1980?

(A) $200,000  (B) $300,000  (C) $400,000  (D) $500,000  (E) **$600,000**

*For Exercises 31-35 use the graph at the right, which shows the number of Tokyo subway passengers, in millions, during one work-week.*

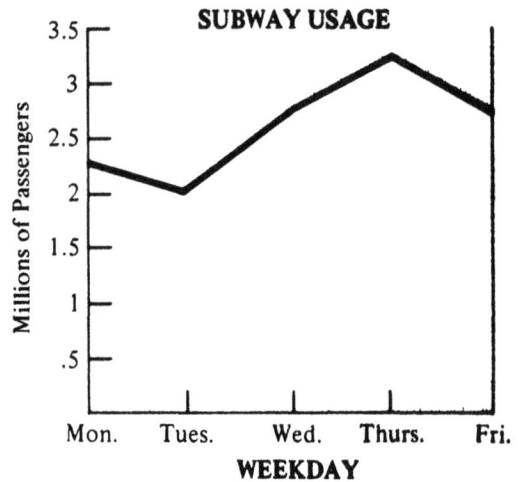

**SUBWAY USAGE**

**31.** The number of passengers on Wednesday was closest to

(A) 2,500,000            (B) 2,750,000              (C) 3,000,000

(D) 3,250,000            (E) 3,500,000

**32.** The smallest number of passengers used the subway on

(A) Monday    (B) Tuesday    (C) Wednesday  (D) Thursday    (E) Friday

**33.** The increase from Tuesday to Thursday was

(A) 750,000            (B) 1,000,000              (C) 1,250,000

(D) 1,500,000            (E) 3,250,000

**34.** The decrease from Monday to Tuesday was

(A) 250,000            (B) 500,000              (C) 1,000,000

(D) 2,000,000            (E) 2,500,000

**35.** The decrease from Thursday to Friday was

(A) 250,000            (B) 500,000              (C) 750,000

(D) 2,750,000            (E) 3,250,000

# 18.  Arithmetic with Negative Integers

An understanding of negative numbers, and how to add, subtract, multiply, and divide them is essential in most subsequent work in algebra and scientific studies.

## A. NEGATIVE NUMBERS AND ABSOLUTE VALUE

*Negative numbers* express such notions as being *below* 0 or *losing* money.

**DEFINITION**

> The numbers $-1, -2, -3, -4, \ldots$ are known as **negative integers**. Other **negative numbers** are $-\frac{1}{2}$, $-.3$, $-.75$, and so on.

Thus

$-5°F.$    describes the temperature (in Fahrenheit) on a cold day

$-\$10$    describes a loss of ten dollars

Negative numbers allow you to subtract a larger integer from a smaller one. For example,

$3 - 5 = -2$

We will now develop the basic rules for arithmetic with negative numbers.

DEFINITION

> It is useful to say that the numbers 5 and −5 are **opposites**. Thus −5 is the opposite of 5, and 5 is the opposite of −5. Also, let 0 be its own opposite.

We will use

$$-a$$

to indicate the opposite of any number $a$—positive, or negative, or 0. Thus

$$\overline{\phantom{x}}\,(-5) = 5$$
$$\uparrow$$
opposite

(This says that the opposite of −5 is 5.)

DEFINITION

> The **sign** of a positive number is + (*plus*);
> The **sign** of a negative number is − (*minus*).

Thus the sign of 5 is  +  ;    the sign of −5 is  −  .

Sometimes we want to ignore the sign of a negative number. For this purpose we introduce **absolute value**.

DEFINITION

> The **absolute value** of a *positive* number or of 0 is the number itself, whereas the absolute value of a negative number is its opposite, that is, the corresponding positive number.

Write

$$|a|$$    for the absolute value of the number $a$

Thus

$|7| = 7$    (The absolute value of the *positive* number 7 is 7 itself.)

$|0| = 0$    (The absolute value of 0 is 0 itself.)

$|-7| = 7$    (The absolute value of the *negative* number −7 is its opposite, 7.)

**Example 1** ▶   Find  (a) $|13|$    (b) $|-12|$    (c) $|-.8|$

*Solution.*

(a)  13 is positive. Thus $|13| = 13$

(b)  −12 is negative. Thus $|-12| = 12$. (The absolute value of −12 is its opposite, 12.)

(c)  −.8 is negative. Thus $|-.8| = .8$                ◀

*Try the exercises for Topic A on page 127.*

## B. ADDITION WITH NEGATIVE NUMBERS

> *To add two negative numbers:*
>
> 1. Add their absolute values.
> 2. Prefix their sum with a minus sign.

For example,

$$(-6) + (-3) = -(6 + 3) = -9$$

**Example 2 ▶** Add.    $(-7) + (-5)$

*Solution.*

1. Add their absolute values, 7 and 5.

    $7 + 5 = 12$

2. Prefix their sum with a minus sign. Thus

    $$(-7) + (-5) = -(7 + 5) = -12$$
    $$\llcorner 12 \lrcorner$$
    ◀

For any number $a$,

> $$a + 0 = 0 + a = a$$
> $$a + (-a) = (-a) + a = 0$$

Thus

$$6 + 0 = 6$$
$$0 + (-8) = -8$$
$$5 + (-5) = 0$$
$$(-3) + 3 = 0$$

> *To add a positive number p and a negative number, other than $-p$:*
>
> 1. Subtract the smaller absolute value from the larger absolute value.
> 2. Prefix the difference with the sign of the *larger* number in absolute value.

**Example 3 ▶** Add.    (a) $8 + (-5)$    (b) $3 + (-9)$

**Solution.**

(a)  $|8| > |-5|$
       ⌊8⌋     ⌊5⌋

Here the *positive* number, 8, has the larger absolute value. Prefix the difference, 8 – 5, with the sign of 8, which is +. (Usually, + is *not* written.)

$$8 + (-5) = 8 - 5 = 3$$

(b)  $|-9| > |3|$
       ⌊9⌋     ⌊3⌋

Here the negative number, -9, has the larger absolute value. Prefix the difference, 9– 3, with the sign of -9, which is -.

$$3 + (-9) = -(9 - 3) = -6 \qquad \blacktriangleleft$$
                 ⌊6⌋

When adding several numbers—some positive, some negative—the following rules of addition are important. Let $a, b$, and $c$ be any numbers. Then

| | |
|---|---|
| $a + b = b + a$ | **(Commutative Law of Addition)** |
| $(a + b) + c = a + (b + c)$ | **(Associative Law of Addition)** |

For example,

$$6 + 3 = 3 + 6 = 9 \qquad \text{by the commutative law}$$

$$(5 + 2) + 4 = 5 + (2 + 4) = 11 \qquad \text{by the associative law}$$
   ⌊7⌋                    ⌊6⌋

The commutative and associative laws enable you to rearrange numbers when adding them. Thus

$$4 + (-8) + 3 = 4 + [(-8) + 3]$$
$$= 4 + [3 + (-8)] \qquad \text{by the commutative law}$$
$$= [4 + 3] + (-8) \qquad \text{by the associative law}$$
$$= 7 + (-8)$$
$$= -(8 - 7)$$
$$= -1$$

*To add three or more numbers*, use the commutative and associative laws.

1. Rearrange, and add the positive numbers and the negative numbers separately.

2. Let $p$ be the sum of the positive numbers, and let $-n$ be the sum of the negative numbers. Then the total sum is $p + (-n)$, or $p - n$.

**Example 4 ▶**    Add.      17
                              14
                             -18
                              23
                             -19

*Solution.* Rearrange, and add the positive numbers and the negative numbers separately.

| 17 | -18 | 54 |
|----|-----|-----|
| 14 | -19 | -37 |
| 23 | -37 | 17 |
| 54 | | |

The resulting sum, 54 + (-37), equals 17.                                      ◀

*Try the exercises for Topic B on page 127.*

## C. SUBTRACTION WITH NEGATIVE NUMBERS

**DEFINITION**

> Let $a$ and $b$ be *any* two numbers—positive, negative, or 0. Define
>
> $$a - b = a + (-b)$$

Thus *subtracting b yields the same result as adding -b.* For example,

$$4 - 6 = 4 + (-6)$$
$$= -(6 - 4)$$
$$= -2$$

Furthermore, it follows from the above definition that

> $$a - 0 = a + (-0) = a + 0 = a$$
> $$0 - a = 0 + (-a) = -a$$

**Example 5 ▶**    Subtract.      (a)  5 - (-4)        (b)  -5 - (-4)

*Solution.* Recall that the opposite of -4 is 4.

$$-(-4) = 4$$

(a)  $5 - (-4) = 5 + 4 = 9$

(b)  $-5 - (-4) = -5 + 4$
$$= -(5 - 4)$$
$$= -1$$                                                          ◀

*Try the exercises for Topic C on page 127.*

## D. MULTIPLICATION WITH NEGATIVE NUMBERS

When the factors of a product are denoted by letters, multiplication is often indicated by a *dot* rather than by a *cross*. Thus

$$a \cdot b \qquad \text{means} \qquad a \times b$$

The commutative and associative laws also hold for multiplication, and enable you to rearrange the factors. Let $a$, $b$, and $c$ be any numbers. Then

| | |
|---|---|
| $a \cdot b = b \cdot a$ | **Commutative Law of Multiplication** |
| $(a \cdot b)c = a(b \cdot c)$ | **Associative Law of Multiplication** |

Furthermore,

$$a \cdot 1 = 1 \cdot a = a$$

For example,

$$4 \times 5 = 5 \times 4 = 20 \qquad \text{by the commutative law}$$

$$\underbrace{(4 \times 5)}_{20} \times 2 = 4 \times \underbrace{(5 \times 2)}_{10} = 40 \qquad \text{by the associative law}$$

and, of course,

$$4 \times 1 = 1 \times 4 = 4$$

Next, here are the rules for determining *the sign of a product*:

If $a$ and $b$ are both positive, then

$$a \cdot (-b) = (-a) \cdot b = - (a \cdot b)$$

$$(-a) \cdot (-b) = a \cdot b$$

$$a \cdot 0 = (-a) \cdot 0 = 0$$

Thus
$$2 \times (-3) = (-2) \times 3 = -(2 \times 3) = -6$$
$$(-2) \times (-3) = 6$$
$$2 \times 0 = (-2) \times 0 = 0$$

In other words, for *nonzero* factors:

1. when 1 factor is negative, the product is *negative*;
2. when 2 factors are negative, the product is *positive*.

More generally,

3. the product of an *odd* number (1, 3, 5, and so on) of negative factors is *negative*;
4. the product of an *even* number (2, 4, 6, and so on) of negative factors is *positive*.

Often, the multiplication symbol is omitted before parentheses. For example, write

$$4(-3) \qquad \text{instead of} \qquad 4 \times (-3)$$

Here *the parentheses indicate multiplication* and *not* subtraction. Thus

$$4(-3) = -(4 \times 3) = -12$$

whereas

$$4 - 3 = 1$$

**Example 6 ▶**  Multiply.      (a) $7(-4)$      (b) $(-5)(-8)$

*Solution.*

(a)  $7(-4) = -(7 \times 4) = -28$

There is one negative factor. The product is negative.

(b)  $(-5)(-8) = 5 \times 8 = 40$

There are two negative factors. The product is positive.  ◀

*Try the exercises for Topic D on page 127.*

## E.  DIVISION WITH NEGATIVE NUMBERS

Instead of writing $a \div b$ , we often write $\frac{a}{b}$.

---

If $a$ and $b$ are both positive, then

$$\frac{-a}{b} = \frac{a}{-b} = -\frac{a}{b} \qquad (negative)$$

$$\frac{-a}{-b} = \frac{a}{b} \qquad (positive)$$

Thus, if the numerator and denominator have the *same sign*, the quotient is *positive*. If the numerator and denominator have *different signs*, the quotient is *negative*. Furthermore,

$$\frac{a}{a} = 1$$

$$\frac{a}{1} = a$$

$$\frac{a}{-1} = -a$$

$$\frac{0}{b} = \frac{0}{-b} = 0$$

and finally,

$$\frac{a}{0} \quad \text{and} \quad \frac{-a}{0} \quad \text{are } undefined.$$

---

Thus 0 *divided by* any *nonzero* number yields 0. But division *by* 0 is undefined (*not* defined).

**Example 7 ▶** Divide.　　(a) $\dfrac{-8}{4}$　　(b) $\dfrac{12}{-3}$　　(c) $\dfrac{-10}{-2}$

***Solution.***

(a) Numerator and denominator have different signs. The quotient is negative.

$$\frac{-8}{4} = -\frac{8}{4} = -2$$

(b) Numerator and denominator have different signs. The quotient is negative.

$$\frac{12}{-3} = -\frac{12}{3} = -4$$

(c) Numerator and denominator have the same sign. The quotient is positive.

$$\frac{-10}{-2} = \frac{10}{2} = 5$$　　　　　◀

**Example 8 ▶** Divide, or indicate that division is undefined.

(a) $\dfrac{0}{4}$　　(b) $\dfrac{0}{-4}$　　(c) $\dfrac{-4}{0}$　　(d) $\dfrac{0}{0}$

***Solution.*** In (a) and (b) use the fact that 0 divided by any *nonzero* number yields 0.

(a) $\dfrac{0}{4} = 0$　　　　(b) $\dfrac{0}{-4} = 0$

In (c) and (d) use the fact that division by 0 is undefined.

(c) $\dfrac{-4}{0}$ is *undefined.*　　　　(d) $\dfrac{0}{0}$ is *undefined.*

Note that even though the numerator is 0, division by 0 is *not* defined.　　◀

*Try the exercises for Topic E on page 128.*

## F.　APPLICATIONS

**Example 9 ▶** The temperature in Vancouver rises from $-9°$ Fahrenheit to $-2°$ Fahrenheit. How many degrees does it rise?

***Solution.*** Subtract $-9$ from $-2$.

$$-2 - (-9) = -2 + 9 = 7$$

The temperature rises $7°$ Fahrenheit.　　　　◀

*Try the exercises for Topic F on page 128.*

## EXERCISES

**A.** *Find each absolute value.*

1. $|5|$       *Answer:*       2. $|-5|$       *Answer:*

3. $|-15|$       *Answer:*       4. $|0|$       *Answer:*

5. $|.73|$       *Answer:*       6. $|-.73|$       *Answer:*

7. $\left|\dfrac{1}{3}\right|$       *Answer:*       8. $\left|-\dfrac{1}{3}\right|$       *Answer:*

**B.** *Add.*

9. $(-3) + (-2)$       *Answer:*       10. $(-8) + (-6)$       *Answer:*

11. $8 + (-6)$       *Answer:*       12. $(-8) + 6$       *Answer:*

13. $(-3) + 7$       *Answer:*       14. $9 + (-5)$       *Answer:*

15. $(-10) + (-10)$       *Answer:*       16. $-10 + 10$       *Answer:*

17. $4 + (-8)$       *Answer:*       18. $8 + (-8)$       *Answer:*

19. $9 + (-12)$       *Answer:*       20. $(-15) + (-7)$       *Answer:*

| 21. | 22. | 23. | 24. |
|---|---|---|---|
| 12 | $-34$ | 52 | 28 |
| $-9$ | $-18$ | $-28$ | $-53$ |
| 15 | 19 | $-37$ | 41 |
| $-17$ | 26 | 41 | $-14$ |
| | $-7$ | $-16$ | 12 |

**C.** *Subtract.*

25. $4 - 5$       *Answer:*       26. $8 - 10$       *Answer:*

27. $6 - (-1)$       *Answer:*       28. $-6 - (-1)$       *Answer:*

29. $(-6) - 1$       *Answer:*       30. $7 - (-2)$       *Answer:*

31. $-10 - (-3)$       *Answer:*       32. $9 - (-6)$       *Answer:*

33. $12 - (-11)$       *Answer:*       34. $(-12) - 11$       *Answer:*

35. $-9 - (-9)$       *Answer:*       36. $0 - (-4)$       *Answer:*

**D.** *Multiply.*

37. $6 (-2)$       *Answer:*       38. $(-10) 5$       *Answer:*

**39.** (–3) (–4)    *Answer:*        **40.** 8 (–5)    *Answer:*

**41.** (–7) (–5)    *Answer:*        **42.** 7 (–5)    *Answer:*

**43.** (–6) (–6)    *Answer:*        **44.** 12 (–3)    *Answer:*

**45.** (–4) 0    *Answer:*        **46.** 0 · 0    *Answer:*

**47.** (–8) (–12)    *Answer:*        **48.** (–9) 11    *Answer:*

**E.** *Divide, or indicate that division is undefined.*

**49.** $\frac{-6}{3}$    *Answer:*        **50.** $\frac{6}{-3}$    *Answer:*

**51.** $\frac{-6}{-3}$    *Answer:*        **52.** $\frac{15}{-5}$    *Answer:*

**53.** $\frac{-20}{4}$    *Answer:*        **54.** $\frac{27}{-3}$    *Answer:*

**55.** $\frac{32}{-8}$    *Answer:*        **56.** $\frac{-63}{7}$    *Answer:*

**57.** $\frac{6}{0}$    *Answer:*        **58.** $\frac{0}{6}$    *Answer:*

**59.** $\frac{-6}{0}$    *Answer:*        **60.** $\frac{0}{-6}$    *Answer:*

**61.** $\frac{0}{0}$    *Answer:*        **62.** $\frac{-12}{-12}$    *Answer:*

**F.**

**63.** The temperature at 8 a.m. in Burlington is –12° Fahrenheit. If the temperature drops 4 degrees the next hour, what is the reading at 9 a.m.?    *Answer:*

**64.** The temperature in Buffalo at 6 p.m. is 2° Celsius. By midnight the temperature has dropped 15 degrees. What is the reading at midnight?    *Answer:*

**65.** A bus travels 12 miles east and then 15 miles west. How far from its starting point is the bus? Is it east or west of the starting point?    *Answer:*

**66.** The temperature change over a 4-hour period is –20 degrees Celsius. Assuming a constant rate of change, how many degrees Celsius does it change each hour?    *Answer:*

# 19.  Several Operations

Frequently, you will have to work with several arithmetic operations — addition, subtraction, multiplication, and division — in the same example. Often these examples involve *squaring* a number. This will be discussed first.

## A.  SQUARING A NUMBER

Recall that

$$a^2 = a \cdot a$$

Thus

$$6^2 = 6 \times 6 = 36$$

and

$$0^2 = 0 \times 0 = 0$$

If $p$ is positive, then $-p$ is negative, and

$$(-p)^2 = (-p)(-p) = p \times p = p^2$$

The square of a negative number is positive because the product of two negative factors is positive. For example,

$$(-4)^2 = (-4)(-4) = 4 \times 4 = 16$$

**Example 1** ▶  Find:      (a) $(-7)^2$      (b) $(-8)^2$

***Solution.***

(a)  $(-7)^2 = (-7)(-7) = 7 \times 7 = 49$

(b)  $(-8)^2 = (-8)(-8) = 8 \times 8 = 64$

◀

*Try the exercises for Topic A on page 132.*

## B. HIGHER POWERS

For any number $a$,

$a^3$    means    $a \cdot a \cdot a$    (with 3 factors)

$a^4$    means    $a \cdot a \cdot a \cdot a$    (with 4 factors)

and so on. Thus

$$2^3 = 2 \times 2 \times 2 = 8$$

$$2^4 = 2 \times 2 \times 2 \times 2 = 16$$

**Example 2** ▶ Find:    (a) $10^3$    (b) $10^4$

*Solution.*

(a)  $10^3 = 10 \times 10 \times 10 = 1000$

(b)  $10^4 = 10 \times 10 \times 10 \times 10 = 10,000$    ◀

**DEFINITION**

In the expression $10^3$, 10 is called the **base** and 3 the **exponent**. We speak of $10^3$ as **raising 10 to the third power**.

*Try the exercises for Topic B on page 133.*

## C. ORDER OF OPERATIONS

*By convention*, the expression

$4 \times 3^2$    means    4 times the square of 3

Thus

$$4 \times 3^2 = 4 \times 9 = 36$$

Here the exponent 2 refers only to the base 3. If you want to indicate the square of the product of 4 and 3, you must use parentheses. Thus

$$(4 \times 3)^2 = 12^2 = 144$$

Here the exponent applies to both factors 4 and 3.

The expression

$5 + 4 \times 2$    means    add the product of 4 and 2 to 5

Thus

$$5 + 4 \times 2 = 5 + 8 = 13$$

If the sum of 5 and 4 is to be multiplied by 2, use parentheses. Thus in the expression

$$(5 + 4)2$$

first add within the parentheses.

$$(5 + 4)2 = 9 \times 2 = 18$$

In general, the order in which addition, subtraction, multiplication, division, and raising to a power are applied in an example is often crucial. If parentheses are given, *first perform the operations within parentheses*. Otherwise:

1. First raise to a power.
2. Then multiply or divide from left to right.
3. Then add or subtract from left to right.

To remember the order of operations think of:

$\boxed{R}$est $\boxed{p}$lease, $\boxed{m}$y $\boxed{d}$ear $\boxed{A}$unt $\boxed{S}$ally

$\boxed{R}$aise to a $\boxed{p}$ower; $\boxed{m}$ultiply or $\boxed{d}$ivide; $\boxed{a}$dd or $\boxed{s}$ubtract

**Example 3** ▶ $4(6 - 3) =$

(A) 24      (B) 21      (C) 12      (D) -12      (E) -21

*Solution.* Here parentheses are given. First perform the operation within parentheses.

$$4(6 - 3) = 4(3) = 12$$

The correct choice is (C). ◀

**Example 4** ▶ $3 + 2 \times 5^2 =$

(A) 53      (B) 125      (C) 103      (D) 169      (E) 13

*Solution.* First raise to a power (*square*); then multiply; then add.

$$3 + 2 \times 5^2 = 3 + 2 \times 25$$
$$= 3 + 50$$
$$= 53$$

The correct choice is (A). ◀

**Example 5 ▶**    Find   $(-3)^2 + 4(-5)$.

> **Solution.** First raise to a power; then multiply. (In this particular example, these operations can be done in the same step.) Then add.
>
> $$(-3)^2 + 4(-5) = 9 + (-20) = -11$$
> $$\underbrace{\phantom{}}_{9} \qquad \underbrace{\phantom{}}_{-20}$$                                                   ◀

**Example 6 ▶**    Find   $2(-4)^2 + 3(-7)$.

> **Solution.** First raise to a power; then multiply; then add.
>
> $$2(-4)^2 + 3(-7) = 2(16) + 3(-7)$$
> $$\underbrace{\phantom{}}_{16} \qquad\qquad \underbrace{\phantom{}}_{32}\ \underbrace{\phantom{}}_{-21}$$
> $$= 32 + (-21)$$
> $$= 11$$                                                   ◀

*Try the exercises for Topic C on page 133.*

## D.   APPLICATIONS

**Example 7 ▶**    A woman buys six 2-dollar glasses and four 5-dollar frying pans. She pays for her purchase with a $50 bill. How much change does she receive?

> **Solution.**
>
> $$50 - \left(6(2) + 4(5)\right) = 50 - (12 + 20)$$
> $$= 50 - 32$$
> $$= 18$$
>
> Her purchases amount to $32, so that she receives $18 change.                                                   ◀

*Try the exercises for Topic D on page 134.*

## EXERCISES

**A.**  *Find each square.*

1. $(-2)^2$        *Answer:*        2. $(-3)^2$        *Answer:*

3. $(-6)^2$        *Answer:*        4. $(-9)^2$        *Answer:*

5. $(-10)^2$        *Answer:*        6. $(-1)^2$        *Answer:*

7. $(-11)^2$        *Answer:*        8. $(-12)^2$        *Answer:*

9. $0^2$        *Answer:*        10. $(-100)^2$        *Answer:*

**B.** *Find each power.*

11.  $3^3$          *Answer:*          12.  $3^4$          *Answer:*

13.  $4^3$          *Answer:*          14.  $5^3$          *Answer:*

15.  $1^4$          *Answer:*          16.  $10^5$          *Answer:*

**C.**  *When several choices are given, draw a circle around the correct letter.*

17.  $5(3 + 1) =$

     (A)  16        (B)  20        (C)  9        (D)  54        (E)  25

18.  $7(5 - 2) =$

     (A)  35        (B)  33        (C)  73        (D)  21        (E)  −21

19.  $10(4 - 6) =$

     (A)  34        (B)  46        (C)  20        (D)  −20        (E)  8

20.  $-4(9 - 4) =$

     (A)  −20        (B)  −9        (C)  1        (D)  −40        (E)  −1

21.  $-7(3 - 10) =$

22.  $2 + 5(1 + 8) =$

23.  $1 - 4(7 - 2) =$

24.  $5 - 2(-10) =$

25.  $2 + 4(3 + 7) =$

     (A)  21        (B)  38        (C)  42        (D)  80        (E)  60

26.  $5(3 + 2) + 2(3 - 1) =$

     (A)  22        (B)  21        (C)  30        (D)  29        (E)  38

27.  $2 \cdot 5^2 =$

     (A)  50        (B)  100        (C)  49        (D)  20        (E)  64

28.  $5(-3)^2 =$

     (A)  4        (B)  −4        (C)  45        (D)  −45        (E)  −30

**29.** $1 - 2(-10)^2 =$

    (A) $-201$    (B) $-199$    (C) $201$    (D) $401$    (E) $-399$

**30.** $(-5)^2 + 2(-6) =$

    (A) $13$    (B) $-37$    (C) $21$    (D) $-29$    (E) $-13$

**31.** $4^2 - 3(-2) =$

**32.** $10^2 - (-9)^2 =$

**33.** $2(-4)^2 + 3(-6) =$

**34.** $5^2 - 2(-1)^2 =$

**35.** $(4^2 - 1)5 =$

**36.** $(2 + 5)^2 - 1 =$

**37.** $2 + 5^2 - 1 =$

**38.** $2^2 + 5^2 - 1 =$

**39.** $(-4)5 - (-4)^2 =$

**40.** $(-3)(-2) + (-5)^2 =$

**41.** $(-2)(-5)^2 - 5(-2)^2 =$

**42.** $7(2 - 4)^2 - 3 =$

**43.** $(-4)(3 - 5)^2 - 2(1 - 2) =$

**44.** $(5 - 8)^2 - (3 + 2)^2 =$

**D.**

  **45.** A bookstore sells 10 books at $6 each and 5 books at $8 each. How much money does it receive?   *Answer:*

  **46.** A theater sells 42 orchestra seats at $10, 58 mezzanine seats at $8, and 102 balcony seats at $6. How much money does it receive?   *Answer:*

  **47.** Ann reads 40 pages per hour. If she has read 135 pages of a 375 page book, how long will it take her to finish it?   *Answer:*

  **48.** A 300-pound man wishes to reduce to 220 pounds. If he is to lose 4 pounds per week, for how many weeks must he diet?   *Answer:*

  **49.** A woman buys five pounds of nuts at $3 per pound, six pounds of cookies at $5 per pound, and three quarts of ice cream at $2 per quart. If she pays for these items with a $100 bill, how much change does she receive?   *Answer:*

**50.** A business makes a profit of $100,000. Of this, $78,000 is reinvested. If the rest of the money is split equally among four partners, how much is each partner's share? *Answer:*

**51.** A football team must gain 10 yards to obtain a first down. In three plays it gains 7 yards, loses 12 yards, and gains 11 yards. How many yards does it need for a first down?    *Answer:*

**52.** A typist charges $3 per page. If she types 5 pages per hour, how long will it take her to earn $120?    *Answer:*

# Exam A on Arithmetic

*When several choices are given, draw a circle around the correct letter.*

**1.** Two hundred twenty thousand five hundred is written

(A) 220,000.05          (B) 225,000          (C) 220,500

(D) 220,005          (E) 225,000.05

**2.** $390 - 72 =$

(A) 218          (B) 318          (C) 328          (D) 462          (E) 228

**3.** $5616 \div 18 =$

(A) 31          (B) 31.2          (C) 312          (D) 3120          (E) 321

**4.** The average of 59, 64, 67, and 74 is _____

**5.** A drama group sells 616 tickets to a play at $5 each. The group rents an auditorium for $750 and it has $1200 in other expenses. Its profit is

(A) $1130          (B) $130          (C) $1081          (D) $2080          (E) $3530

**6.** Which of these fractions is the smallest?

(A) $\frac{3}{4}$          (B) $\frac{3}{5}$          (C) $\frac{4}{5}$          (D) $\frac{4}{7}$          (E) $\frac{5}{7}$

**7.**    10 hours 30 minutes
    $-$ 3 hours 45 minutes
    _____

**8.** $\frac{2}{5} + \frac{1}{9} =$          **9.** $4\frac{1}{5} - 2\frac{1}{2} =$          **10.** $3\frac{1}{4} \div 1\frac{3}{4} =$

11.  $12.185 + 10.092 + 4 =$

12.  Which of these numbers is the smallest?

   (A) .071        (B) .008        (C) .069        (D) .0079        (E) .017

13.  $27.1 - 12.45 =$

14.  Change $\frac{10}{13}$ to a decimal rounded to the nearest hundredth.

   (A) .76        (B) .77        (C) .769        (D) .8        (E) .80

15.  If a pound of coffee costs $2.95, how much do 12 pounds of coffee cost?
    *Answer:*

16.  What is 60% of 75?   *Answer:*

17.  If 60 is 75% of a certain number, find that number.   *Answer:*

18.  Find the cost of the material for a 5-foot-by-3-foot tablecloth at $4 per square foot.
    *Answer:*

19.  *Use the graph at the right to find the total number of magazines sold by the Weekly Journal from June through August, 1982.*

   (A) 5500        (B) 7000

   (C) 7500        (D) 8000

   (E) 7.5

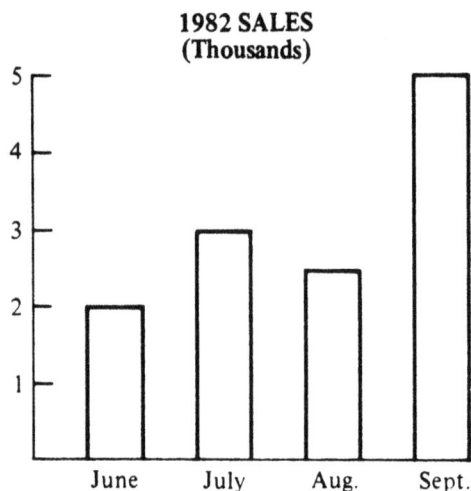

1982 SALES
(Thousands)

20.  $4(-2)^2 + 3(-5) =$

## Exam B on Arithmetic

*When several choices are given, draw a circle around the correct letter.*

1.  Five thousand and two tenths is written

   (A) 5210        (B) 5002.10        (C) 5000.2        (D) 5000.02        (E) 5002.1

2.  $1855 \div 35 =$

3.  A student has scores of 80, 95, 90, and 91 on her math tests. Her average is

   (A) 88        (B) 89        (C) 90        (D) 91        (E) 92

4. A baseball game started at 7:30 P.M. and ended at 10:15 P.M. How long did it last?

   (A) 3 hours 15 minutes    (B) 2 hours 30 minutes    (C) 2 hours 45 minutes

   (D) 3 hours 30 minutes    (E) 3 hours 45 minutes

5. $\frac{1}{10} + \frac{3}{5} =$                    6. $4\frac{7}{8} - 1\frac{1}{4} =$

7. Which of the following fractions is the *smallest*?

   (A) $\frac{1}{4}$        (B) $\frac{2}{5}$        (C) $\frac{3}{10}$        (D) $\frac{5}{12}$        (E) $\frac{3}{13}$

8. Which of the following numbers is the *smallest*?

   (A) .045        (B) .0054        (C) .4005        (D) .0045        (E) .004 49

9. $10 \div 3\frac{1}{4} =$

10. $4.5 \times .48 =$

   (A) 216        (B) 21.6        (C) 2.16        (D) .216        (E) .0216

11. If pencils sell for $.12 each, how many can you buy for $18?

   (A) 216        (B) 15        (C) 150        (D) 1500        (E) 2160

12. $27.9 - 4.95 =$

13. What is 88% expressed as a fraction?

   (A) $\frac{8}{10}$        (B) $\frac{12}{25}$        (C) $\frac{22}{25}$        (D) $1\frac{8}{10}$        (E) 88

14. A storage company charges $100 to store a full-sized piano for the first month and $40 for each additional month. How much does it cost to store this piano for six months?

15. 30% of the students at River Junction College are freshmen. If there are 900 freshmen at this college, how many students are there at the college? *Answer:*

16. If 70% of a number is 7, find this number. *Answer:*

17. At a sale, a $16 sweater is reduced by 25%. Its sale price is

   (A) $4        (B) $20        (C) $12        (D) $15.75        (E) $15.60

18. Find the area of a rectangular movie screen that is 15 feet high and 30 feet wide.

   (A) 2 square feet        (B) 200 square feet        (C) 45 square feet

   (D) 450 square feet        (E) 4500 square feet

**19.** In the graph at the right find
the annual expenditure for
housing.

(A)  $1,920,000      (B)  $1,600,000

(C)  $1,880,000      (D)  $4,000,000

(E)  $19,200,000

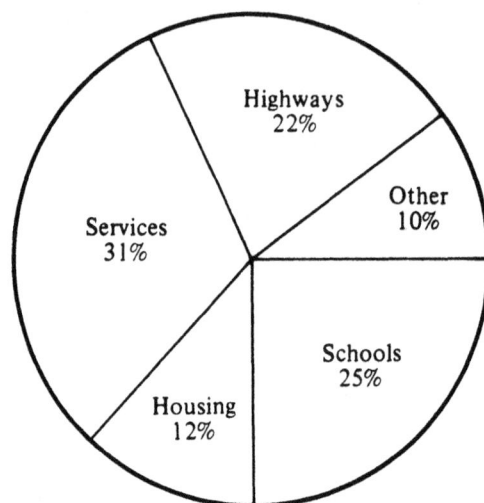

Services 31%

Highways 22%

Other 10%

Schools 25%

Housing 12%

**ANNUAL EXPENDITURES: $16,000,000**

**20.**  $(-5)^2 + 4(-3) =$

# Part Two
# **A**lgebra

---

## 20.  Algebraic Expressions

In this section you will be introduced to some of the fundamental algebraic terminology that will be used throughout the rest of the book. You will first learn how to express certain numerical concepts in terms of algebraic expressions.

### A. VARIABLES

Sometimes, instead of adding, subtracting, multiplying or dividing *individual* numbers, the same arithmetic process can apply to *any number* under discussion. It is important to have symbols that can stand for any one of several numbers. For example, if you want to add 5 to a number (whether that number is 1, 2, 10, or any other number), you can express this by

$x + 5$

If you want to multiply a number by 3, you can write

$3x$

In each case *x* stands for *any* number, and is called a *variable*.

DEFINITION

> **A variable** is a symbol that stands for any number being discussed.

Letters such as *x, y, z, a, b,* and *c* are frequently used as variables. Algebraic expressions such as

$$x + z$$

$$x - y$$

$$4ab$$

$$x^2 + 3x$$

are built up from variables.

Here is how we translate words to symbols.

**Example 1** ▶ Suppose that *x* stands for some number. Express each of the following in terms of *x*.

(a)  4 more than the number

(b)  1 less than the number

(c)  7 times the number

(d)  the number divided by 5

*Solution.*

(a)  First note that

6 is 4 more than the number 2, and 6 = 2 + 4

Also,

7 is 4 more than the number 3, and 7 = 3 + 4

In general, 4 more than the number means to add 4 to the number. Thus

4 *more than* the number

4      +         x              or $x + 4$

(b)  Observe that

3 is 1 less than the number 4, and 3 = 4 - 1

Also,

4 is 1 less than the number 5, and 4 = 5 - 1

In general, 1 less than the number means to subtract 1 from the number. Thus

$$\text{1 }\textit{less than}\text{ the number} \quad\quad \text{means} \quad\quad \underbrace{\text{the number}}\ \underbrace{\textit{minus}}\ 1$$

$$x \qquad\qquad - \qquad 1$$

(c) $\underbrace{7}\ \underbrace{\text{times}}\ \underbrace{\text{the number}}$

$$7 \quad\cdot\quad x \qquad\qquad \text{or } 7x$$

(d) $\underbrace{\text{the number}}\ \underbrace{\text{divided by}}\ 5$

$$x \qquad \div \qquad 5 \qquad \text{or } \frac{x}{5}$$

◀

**Example 2 ▶** Suppose that $a$ stands for some number. Express each of the following in terms of $a$.

(a) the number increased by 8

(b) the number decreased by 10

(c) one-third of the number

(d) 2 more than one-third of the number

(e) the square of the number

*Solution.*

(a) *Increasing* a number by 8 means *adding* 8 to the number. Thus

$$\underbrace{\text{the number}}\ \underbrace{\text{increased by}}\ 8$$

$$a \qquad\quad + \qquad 8$$

(b) *Decreasing* a number by 10 means *subtracting* 10 from the number. Thus

$$\underbrace{\text{the number}}\ \underbrace{\text{decreased by}}\ 10$$

$$a \qquad\quad - \qquad 10$$

(c) Here "of" indicates multiplication.

$$\underbrace{\text{one-third}}\ \text{of}\ \underbrace{\text{the number}}$$

$$\frac{1}{3} \quad\cdot\quad a \qquad\quad \text{or } \frac{1}{3}a \qquad \text{or } \frac{a}{3}$$

(d) Use part (c).

$$\underbrace{\text{2 more than}}\ \underbrace{\text{one-third of a number}}$$

$$2\ + \qquad\qquad \frac{a}{3} \qquad\qquad \text{or } \frac{a}{3} + 2$$

(e) the square of the number
$$\overbrace{\phantom{\text{the square of the number}}}$$
$$\ddot{a}^2$$

*Try the exercises for Topic A on page 143.*

## B.  APPLICATIONS

**Example 3** ▶   Find the value, in cents, of *n* nickels and *q* quarters.

(A) $n + q$     (B) $n - q$     (C) $125nq$     (D) $5n + 25q$     (E) $5(n + 25q)$

*Solution.*  One nickel is worth 5 cents. Thus

   *n* nickels are worth $n \times 5$ cents,     or     $5n$ cents

One quarter is worth 25 cents. Thus

   *q* quarters are worth $q \times 25$ cents,     or     $25q$ cents

*Together*, the value in cents is as indicated:

$$\underbrace{n \text{ nickels and } q \text{ quarters}}$$
$$5n \quad + \quad 25q$$

The correct choice is (D).   ◀

   If a car travels 50 miles per hour, then in 2 hours it travels $50 \times 2$, or 100, miles. Similarly, if a boy walks 3 miles per hour, then in 4 hours he walks $3 \times 4$, or 12, hours. In general, when an object travels at a *constant* rate, multiply the *rate* of travel by the *time* spent traveling to obtain the *distance* traveled.

$$\boxed{\text{Rate} \;\times\; \text{Time} \;=\; \text{Distance}}$$

**Example 4** ▶   A car travels 40 miles per hour for *y* hours. It then increases its speed to 50 miles per hour for *z* more hours. Express the distance traveled in miles.

(A) $40 + y + 50 + z$        (B) $40y + 50 + z$        (C) $40 + y + 50z$

(D) $40(y + 50z)$        (E) $40y + 50z$

*Solution.*  We can express miles *per* hour by the "fraction," $\dfrac{\text{miles}}{\text{hours}}$.

   Rate $\times$ Time = Distance

   $40 \dfrac{\text{miles}}{\cancel{\text{hours}}} \times y \cancel{\text{hours}} = 40y$ miles

$$50 \frac{\text{miles}}{\text{hours}} \times z \text{ hours} = 50z \text{ miles}$$

The car first travels $40y$ miles (40 miles per hour for $y$ hours) and then an *additional* $50z$ miles (50 miles per hour for $z$ hours). The (*total*) distance it travels, in miles, is then given by choice (E),

$$40y + 50z$$ ◀

*Try the exercises for Topic B on this page.*

## EXERCISES

**A.** *In Exercises 1–14 suppose that x stands for some number. Express each of the following in terms of x.*

1.  the number plus 3                          *Answer:*

2.  the number minus 7                         *Answer:*

3.  6 times the number                         *Answer:*

4.  the number divided by 10                   *Answer:*

5.  2 more than the number                     *Answer:*

6.  2 less than the number                     *Answer:*

7.  the number increased by 12                 *Answer:*

8.  the number decreased by 1                  *Answer:*

9.  the square of the number                   *Answer:*

10. one-half of the number                     *Answer:*

11. 1 more than 3 times the number             *Answer:*

12. 2 less than 5 times the number             *Answer:*

13. 6 less than one-half of the number         *Answer:*

14. 10 more than one-fourth of the number      *Answer:*

**B.** *When several choices are given, draw a circle around the correct letter.*

15. Find the value, in cents, of $n$ nickels and $d$ dimes.

(A) $n + d$          (B) $5n + d$          (C) $n - d$

(D) $5n + 10d$          (E) $5(n + 10d)$

16.  Find the value, in cents, of $d$ dimes and $p$ pennies.

(A) $d + p$          (B) $d - p$          (C) $10d + p$

(D) $d + 10p$        (E) $10(d + p)$

17.  Find the value, in cents, of $q$ quarters and $h$ half-dollars.

(A) $q + \dfrac{h}{2}$        (B) $q + 2h$        (C) $25q + \dfrac{h}{2}$

(D) $25q + 2h$        (E) $25q + 50h$

18.  Find the value, in cents, of $n$ nickels, $d$ dimes, and $q$ quarters.

(A) $n + d + q$          (B) $5n + d + q$          (C) $5n + 10d + 25q$

(D) $5q + 10d + 25n$        (E) $5(n + d + q)$

19.  Find the value, in dollars, of $x$ five-dollar bills and $y$ ten-dollar bills.

*Answer:*

20.  Find the value, in dollars, of $m$ one-dollar bills and $n$ five-dollar bills.

*Answer:*

21.  Find the value, in dollars, of $x$ five-dollar bills, $y$ twenty-dollar bills, and $z$ fifty-dollar bills.   *Answer:*

22.  Find the value, in *dollars*, of $n$ nickels and $h$ half-dollars.   *Answer:*

23.  The cost, in cents, of $x$ chocolate bars at 25 cents per chocolate bar is _____

24.  The cost, in dollars, of $r$ bottles of red wine at 5 dollars per bottle and $w$ bottles of white wine at 6 dollars per bottle is _____

25.  A train travels at 50 miles per hour for $h$ hours. Express the distance it travels in miles.

(A) $50h$      (B) $h + 50$      (C) $h - 50$      (D) $\dfrac{h}{50}$      (E) $\dfrac{50}{h}$

26.  A train travels at $m$ miles per hour for 4 hours. Express the distance it travels in miles.

(A) $m + 4$      (B) $m^4$      (C) $4^m$      (D) $4m$      (E) $\dfrac{m}{4}$

27.  A train travels $m$ miles per hour for $h$ hours. Express the distance it travels in miles.

*Answer:*

28.  A train travels at 60 miles per hour for $h$ hours and then at 50 miles per hour for 2 hours. Express the distance it travels in miles.   *Answer:*

29. A train travels at 80 miles per hour for $x$ hours and then at 60 miles per hour for $y$ hours. Express the distance it travels in miles.    *Answer:*

30. A man travels in a train that is going at 50 miles per hour for $x$ hours. When he leaves the train, he then walks 3 miles further. Express the total distance he has gone in miles.    *Answer:*

31. An envelope contains $x$ 20-cent stamps and $y$ 17-cent stamps. What is the total value, in cents, of these stamps?    *Answer:*

32. A television show has $y$ 3-minute commercials and $z$ 2-minute commercials. How many minutes of commercials are there?    *Answer:*

# 21.  Substituting Numbers for Variables

Before considering how to combine algebraic expressions, let us consider how to evaluate an algebraic expression for given values of the variable or variables.

## A. ONE VARIABLE

Recall that a variable is a symbol that stands for any number being discussed. Thus, in an algebraic expression such as $3x$, the variable $x$ may stand for 2, or for 5, or for $-3$. The value of the expression $3x$ will be different in each case. For example, when $x = 2$,  then  $3x = 3(2) = 6$

when $x = 5$,  then  $3x = 3(5) = 15$

when $x = -3$,  then $3x = 3(-3) = -9$

In general, when you are given numbers that the variables of an algebraic expression represent, you can find the value of the expression by substituting the numbers for the variables.

**Example 1** ▶ Find the value of $5t - 2$ when $t = 4$.

(A) 4          (B) 20          (C) 52          (D) 7          (E) 18

*Solution.* Substitute 4 for $t$ in the expression $5t - 2$ to obtain

$5(4) - 2$

Now find this value.

$5(4) - 2 = 20 - 2 = 18$
$\llcorner 20 \lrcorner$

The correct choice is (E).                                                              ◀

**Example 2** ▶ Find the value of $10(5)^x$ when $x = 2$.

*Solution.* Substitute 2 for $x$ in the expression $10(5)^x$ to obtain

$10(5)^2$

Raise to a power (here, square the 5) before multiplying.

$10(5)^2 = 10(5 \times 5)$
$\phantom{10(}\llcorner 5 \times 5 \lrcorner$
$= 10(25)$
$= 250$                                                              ◀

When a variable occurs more than once in an expression, *each time the variable occurs*, substitute the given number for that variable. Thus if $a = 3$, then the value of the expression

$a^2 + 5a$

is

$3^2 + 5(3) = 9 + 15 = 24$
$\llcorner 9 \lrcorner \quad \llcorner 15 \lrcorner$

**Example 3** ▶ Find the value of $3y^2 - 2y + 1$ when $y = -2$.

(A) 17          (B) -15          (C) 9          (D) -7          (E) 31

*Solution.* Substitute -2 for *each occurrence* of $y$ in the expression

$$3y^2 - 2y + 1$$

to obtain

$$3(-2)^2 - 2(-2) + 1$$

First raise to a power (here, *square* the $-2$), then multiply, then add and subtract.

$$3(-2)^2 - 2(-2) + 1 = 3(4) - 2(-2) + 1$$
$$\underbrace{\phantom{(-2)^2}}_{4} \qquad\qquad \underbrace{\phantom{3(4)}}_{12} \underbrace{\phantom{-2(-2)}}_{-4}$$
$$= 12 - (-4) + 1$$
$$= 12 + 4 + 1$$
$$= 17$$

In the third step, recall that *subtracting* $-4$ yields the same as *adding* 4. The correct choice is (A).  ◀

*Try the exercises for Topic A on page 149.*

## B. TWO VARIABLES

Sometimes the expressions that are to be evaluated contain two or more variables. Substitute each given number for the specified variable.

**Example 4** ▶ Find the value of $x + 2y^2$ when $x = -1$ and $y = 3$.

(A) 2    (B) 17    (C) 35    (D) 25    (E) 37

*Solution.* Substitute $-1$ for $x$ and 3 for $y$ in the expression

$$x + 2y^2$$

Raise to a power, then multiply, then add.

$$-1 + 2(3)^2 = -1 + 2(9) = -1 + 18 = 17$$
$$\underbrace{\phantom{(3)^2}}_{9} \qquad \underbrace{\phantom{2(9)}}_{18}$$

The correct choice is (B).  ◀

**Example 5** ▶ What is the value of $4pq + q^2$ when $p = 2$ and $q = -3$?

*Solution.* Substitute 2 for $p$ and $-3$ for each occurrence of $q$ in the expression

$$4pq + q^2$$

to obtain

$$4(2)(-3) + (-3)^2 = 4(2)(-3) + 9$$
$$\underbrace{\phantom{(-3)^2}}_{9} \quad \underbrace{\phantom{4(2)}}_{8}$$
$$\underbrace{\phantom{4(2)(-3)}}_{-24}$$
$$= -24 + 9$$
$$= -15$$  ◀

**Example 6 ▶**    If $t = 3xy^2$, find $t$ when $x = 2$ and $y = -2$.

> *Solution.* In order to find the value of $t$, substitute 2 for $x$ and $-2$ for $y$ in the expression
>
> $$3xy^2$$
>
> Thus
>
> $$t = 3(2)(-2)^2$$
>
> $$\phantom{t =} = 3(2)(4)$$
>
> $$\phantom{t =} = 24$$    ◀

*Try the exercises for Topic B on page 150.*

## C.  FORMULAS

> Formulas are algebraic equations that indicate the relationship between various quantities.

**Example 7 ▶**    A car travels at the rate of 55 miles per hour. The distance $d$ (in miles) that it travels in $t$ hours is given by the formula

> $$d = 55t$$

Determine the distance it travels  (a) in 2 hours;  (b) in 5 hours.

> *Solution.*
>
> (a)   Substitute 2 for $t$ in the formula $d = 55t$, and obtain
>
> $$d = 55 \cdot 2 = 110$$
>
> Thus in 2 hours the car travels 110 miles.
>
> (b)   Substitute 5 for $t$:
>
> $$d = 55 \cdot 5 = 275$$
>
> In 5 hours the car travels 275 miles.    ◀

**Example 8 ▶**    The area, $A$, of a triangle is given by the formula

> $$A = \frac{bh}{2}$$

where $b$ is the length of the base and $h$ is the length of the altitude. Find the area if (a) $b = 8$ feet and $h = 10$ feet; (b) $b = 20$ feet and $h = 15$ feet. (See the accompanying figure.)

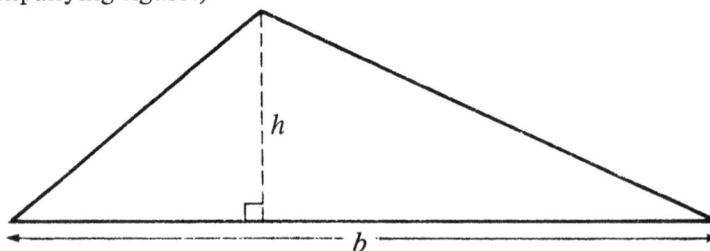

*Solution.*

(a)  Substitute 8 for $b$ and 10 for $h$ in the formula $A = \frac{bh}{2}$, and obtain

$$A = \frac{8 \cdot 10}{2} = 40$$

Thus the area is 40 square feet.

(b)  Substitute 20 for $b$ and 15 for $h$:

$$A = \frac{20 \cdot 15}{2} = 150$$

Thus the area is 150 square feet.    ◀

*Try the exercises for Topic C on page 151.*


## EXERCISES

**A.**  *When several choices are given, draw a circle around the correct letter.*

1.  Find the value of $4x$ when $x = 3$.

  (A) 43          (B) 12          (C) 7          (D) 64          (E) 1

2.  Find the value of $y + 5$ when $y = 2$.

  (A) 2          (B) 7          (C) 25          (D) 32          (E) −3

3.  Find the value of $a - 3$ when $a = 3$.

  (A) 3          (B) −3          (C) 0          (D) −6          (E) −9

4.  Find the value of $b^2 + 1$ when $b = 5$.

  (A) 6          (B) 25          (C) 26          (D) 36          (E) 11

5.  If $m = -3$, what is the value of $4m - 3$?

  (A) 1          (B) −1          (C) −6          (D) −15          (E) 36

6.  If $n = -1$, find the value of $1 - n$.

  (A) 1          (B) −1          (C) 0          (D) 2          (E) −2

7.  Find the value of $4^a$ when $a = 2$.    *Answer:*

8.  Find the value of $10^b$ when $b = 2$.    *Answer:*

9.  Find the value of $-3^x$ when $x = 2$.    *Answer:*

10.  Find the value of $(-5)^y$ when $y = 2$.    *Answer:*

11.  Find the value of $10^m$ when $m = 4$.  *Answer:*

12.  Find the value of $(-1)^n$ when $n = 6$.  *Answer:*

13.  Find the value of $3(2)^x$ when $x = 2$.  *Answer;*

14.  Find the value of $-5(3)^x$ when $x = 2$.  *Answer:*

15.  Find the value of $1 - 3^x$ when $x = 2$.  *Answer:*

16.  Find the value of $(1 - 3)^x$ when $x = 2$.  *Answer:*

17.  Find the value of $x^2 + 6x$ when $x = 2$.

 (A) 10   (B) 12   (C) 14   (D) 16   (E) 20

18.  Find the value of $y^2 - y$ when $y = -1$.

 (A) 1   (B) -1   (C) 2   (D) -2   (E) 0

19.  If $a = 5$, what is the value of $2a^2 - 3a$?  *Answer:*

20.  If $b = -4$, what is the value of $4 - b^2$?  *Answer:*

21.  Evaluate $p^3 - 2p$ when $p = 10$.  *Answer:*

22.  Evaluate $t^4 + 6t + 1$ when $t = -1$.  *Answer:*

**B.** *When several choices are given, draw a circle around the correct letter.*

23.  Find the value of $p + q$ when $p = 5$ and $q = -7$.

 (A) 12   (B) -12   (C) 2   (D) -2   (E) -35

24.  Find the value of $2a + 5b$ when $a = 3$ and $b = 4$.

 (A) 7   (B) 14   (C) 11   (D) 26   (E) 35

25.  Find the value of $2c - 4d$ when $c = 6$ and $d = 4$.

 (A) -18   (B) 4   (C) -4   (D) -10   (E) -48

26.  Find the value of $x + 3y$ when $x = 3$ and $y = -2$.

 (A) 3   (B) 6   (C) 1   (D) 0   (E) -3

27.  Find the value of $5m - 2n$ when $m = -1$ and $n = -2$.  *Answer:*

28.  Find the value of $10 - 2y - 3z$ when $y = 2$ and $z = -1$.  *Answer:*

29.  Find the value of $a^2 + 2b$ when $a = 5$ and $b = 4$.  *Answer:*

30. Find the value of $x - y^2$ when $x = 4$ and $y = 3$.  *Answer:*

31. Find the value of $m + 3n^2$ when $m = 2$ and $n = 5$.  *Answer:*

32. Find the value of $2a - b^2$ when $a = 3$ and $b = -2$.  *Answer:*

33. Find the value of $5s + t^2$ when $s = -2$ and $t = -3$.  *Answer:*

34. Find the value of $x^2 - 2y^2$ when $x = 2$ and $y = -2$.  *Answer:*

35. Find the value of $a^2 + 2ab$ when $a = 2$ and $b = 5$.
    (A) 14    (B) 24    (C) 40    (D) 45    (E) 80

36. Find the value of $p^2 - 4pq$ when $p = -2$ and $q = 3$.
    (A) 24    (B) 28    (C) -15    (D) -20    (E) 0

37. What is the value of $x^2 + 10xy$ when $x = 10$ and $y = -1$?
    (A) 200    (B) 100    (C) 110    (D) 90    (E) 0

38. If $a = 4$ and $b = -2$, find the value of $a^2 + 4ab$.
    (A) 16    (B) 32    (C) 48    (D) -16    (E) 0

39. If $x = -2$ and $y = -3$, find the value of $x^2 - 3y^2$.  *Answer:*

40. If $m = 5$ and $n = 10$, find the value of $m^2 + 2mn + 1$.  *Answer:*

41. Find the value of $2ab + a^2 + b^2$ if $a = 2$ and $b = 3$.  *Answer:*

42. Find the value of $x^2 + 3xy$ when $x = -1$ and $y = 10$.  *Answer:*

43. Find the value of $x^2 + 2xy^2$ when $x = 3$ and $y = 2$.  *Answer:*

44. Find the value of $x^2 + 10xy$ when $x = -10$ and $y = -10$.  *Answer:*

45. If $c = 2ab^2$, find $c$ when $a = 3$ and $b = 2$.  *Answer:*

46. If $z = 5xy^2$, find $z$ when $x = 2$ and $y = -2$.  *Answer:*

47. If $a = 3m^2n$, find $a$ when $m = -1$ and $n = -2$.  *Answer:*

48. If $t = 4x^2y^2$, find $t$ when $x = 3$ and $y = -2$.  *Answer:*

C.

49. Find the area of a square if each side is of length  (a) 7 inches, (b) 12 inches.
    *Answer:*     (a)                     (b)

50. Find the area of a triangle if
    (a) $b = 10$ inches and $h = 17$ inches; (b) $b = 16$ inches and $h = 25$ inches.
    *Answer:*     (a)                     (b)

51. An airplane travels at the rate of 400 miles per hour. The distance $d$ (in miles) that it travels in $t$ hours is given by the formula
    $$d = 400t$$
    Determine the distance it travels  (a) in 3 hours; (b) in 9 hours.
    *Answer:*     (a)                     (b)

**52.** The volume of a rectangular box is given by the formula $V = lwh$, where $l$ is the length, $w$ the width, and $h$ the height. Find the volume if

   (a)   $l = 18$ inches,
          $w = 5$ inches,
          and $h = 6$ inches;

   (b)   $l = 10$ inches,
          $w = 9$ inches,
          and $h = 7$ inches.

  (See the accompanying figure.)

*Answer:*    (a)

              (b)

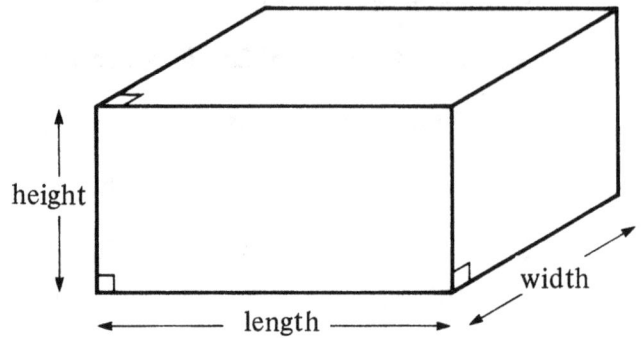

**53.** The surface area of a rectangular box is given by the formula $S = 2lw + 2lh + 2wh$, where $l$, $w$, and $h$ are as in Exercise 52.

Find the surface area if

   (a)   $l = 12$ feet, $w = 5$ feet, and $h = 10$ feet;
   (b)   $l = 15$ feet, $w = 12$ feet, and $h = 6$ feet.

*Answer:*    (a)              (b)

**54.** The inside of a cigar box is of length 11 inches, width 7 inches, and height 5 inches.

   (a)   What is the volume of the inside of this box?
   (b)   How much paper is needed to line the inside? (*See Exercises 52 and 53.*)

*Answer:*    (a)              (b)

**55.** A piece of material cost $7 per square meter. The cost $C$ (in dollars) for $x$ square meters of this material is given by the formula

$$C = 7x$$

Determine the cost of (a) 4 square meters; (b) 7 square meters of this material.

*Answer:*    (a)              (b)

**56.** A ball is thrown upward from ground level with an initial velocity of 96 feet per second. Its height $h$ (in feet) after $t$ seconds is given by the formula

$$h = 96t - 16t^2$$

Determine its height (a) after 1 second; (b) after 2 seconds; (c) after 6 seconds.

*Answer:*    (a)              (b)              (c)

**57.** The total value $v$ (in dollars) of a number of coins is given by the formula

$$v = \frac{5n + 10d + 25q}{100}$$

where $n$, $d$, and $q$ are, respectively, the number of nickels, dimes, and quarters. Find the total value of 10 nickels, 15 dimes, and 7 quarters.    *Answer:*

# 22.  Adding and Subtracting Polynomials

Here we consider how to add and subtract algebraic expressions known as *polynomials*. Before treating this topic, some preliminary notions must be introduced.

## A. TERMS

DEFINITION

> A **term** is a product of numbers and variables. A number, by itself, and a variable, by itself, are each considered to be terms.

For example, each of the following is a term.

| | |
|---|---|
| 4 | A number, by itself, is a term. |
| $x$ | A variable, by itself, is a term. |
| $4x$ | This is the product of 4 and $x$. |
| $5a^2b$ | $5a^2b = 5 \cdot a \cdot a \cdot b$ |

However, $x + 1$ and $a + b$ are *not* terms because the definition of a term does not permit adding a number and a variable, or adding two different variables.

DEFINITION

> The numerical factor (or the product of numerical factors) in a term is called the **numerical coefficient** or simply, the **coefficient**, of the term.

**Example 1 ▶** What is the coefficient of each of the following terms?

(a) $6x$          (b) $-2x^3y$          (c) 20

(d) $(5a) \cdot (2b)$          (e) $a$          (f) $-b$

*Solution.*

(a)  6

(b)  −2        Note that 3 is the *exponent* of $x$ and is *not* the coefficient.

(c)  20        A (numerical) term is its own coefficient.

(d)  10        The product of the numerical factors 5 and 2 is 10.

(e)  1         $a = 1 \cdot a$

(f)  −1        $-b = (-1) \cdot b$        ◄

It will be useful to say that the **sign of a term** is the sign of its coefficient. Thus the sign of $4x$ is $+$ and the sign of $-4x$ is $-$.

*Try the exercises for Topic A on page 158.*

## B.  POLYNOMIALS

DEFINITION

> A **polynomial** is either a term or a sum of terms.

Each of the following is a polynomial.

$2x$                          A term is a polynomial.

$x + 1$                       A sum of terms is a polynomial.

$x^2 + 3x - 5$               Note that $x^2 + 3x - 5 = x^2 + 3x + (-5)$.

$ab + a + b$

Example 2  ►  $x^2 + 2x - 3$ is the sum of the terms

$\boxed{\phantom{xx}}$ , $\boxed{\phantom{xx}}$ , and $\boxed{\phantom{xx}}$ .

*Solution.*  $x^2 + 2x - 3$ is the sum of the terms

$\boxed{x^2}$ , $\boxed{2x}$ , and $\boxed{-3}$ .        ◄

*Try the exercises for Topic B on page 158.*

## C.  LIKE TERMS

Adding $2a + 3a$ is similar to adding 2 apples and 3 apples, whereas adding
$\underbrace{\phantom{2a + 3a}}_{5a}$ $\phantom{xxxxxxxxxxx}$ $\underbrace{\phantom{2 apples and 3 apples}}_{5 \text{ apples}}$
$2a + 3b$ is similar to adding 2 apples and 3 bananas. You can combine terms when adding them only when they are alike in their variables and in the exponent of each variable.

DEFINITION

> Two or more nonzero terms are said to be **like terms** if
>
> 1.  they contain exactly the same variables, and
>
> 2.  each variable has the same exponent in each term.

In other words *like terms can differ only in their numerical coefficients or in the order of their variables.*

Here are some examples of *like terms*:

$10x$ and $2x$

$y^2$ and $-5y^2$

$3ab$ and $2ab$

$x^2y$ and $yx^2$       Note that like terms can differ in the order of their variables.

$5$ and $-4$       Any two nonzero numbers are like terms because neither contains any variables.

Here are some examples of *unlike terms*:

$7p$ and $3q$       They differ in their variables.

$4x$ and $4x^2$       They differ in the exponents of $x$.

$5xy^2$ and $3xy$       They differ in the exponents of $y$.

**Example 3** ▶ Which of the following pairs are like terms?

(a) $5a^2$ and $-2a^2$       (b) $3b^2$ and $2b^3$       (c) $6st$ and $2su$

***Solution.***

(a) *like terms*: Both terms have the same variable $a$, and the same exponent of $a$.

(b) *unlike terms*: They differ in the exponents of $b$.

(c) *unlike terms*: One contains the variable $t$ and the other contains the variable $u$.       ◀

*Try the exercises for Topic C on page 158.*

## D. ADDING AND SUBTRACTING LIKE TERMS

Observe that

$$3(4 + 2) = 3(6) = 18$$
$$\llcorner 6 \lrcorner$$

But we can also combine as follows:

$$3(4 + 2) = 3(4) + 3(2)$$
$$\llcorner 12 \lrcorner \quad \llcorner 6 \lrcorner$$
$$= 12 + 6$$
$$= 18$$

This second method, though longer, illustrates the way *like terms* can be added and subtracted.

Let $a$ and $b$ be any numbers, and let $x$ be a variable. Then

$$ax + bx = (a + b)x$$

and

$$ax - bx = (a - b)x$$

Thus

$$5x + 3x = (5 + 3)x = 8x$$

and

$$5x - 3x = (5 - 3)x = 2x$$

The above rules also apply when $x$ is replaced by a power of a variable, such as $a^2$, or by a product of variables, such as $xy$.

**Example 4 ▶**    Add or subtract:

(a)  $10a^2 + 4a^2$                    (b)  $6xy - 2xy$

*Solution.*  In each case the terms to be combined are *like* terms.

(a)  $10a^2 + 4a^2 = (10 + 4)a^2 = 14a^2$

(b)  $6xy - 2xy = (6 - 2)xy = 4xy$

*Unlike terms* cannot be added or subtracted in this way. For example, $2x + 3y$ cannot be further simplified.                                                      ◀

*Try the exercises for Topic D on page 158.*

## E. ADDING POLYNOMIALS

To add or subtract polynomials, *group like terms together* and add in columns.

**Example 5 ▶**    Add $4x + 5y$ and $4y - 3x$.

(A)  $8x + 2y$              (B)  $7x + 9y$              (C)  $x + 9y$

(D)  $16xy - 15xy$         (E)  $x^2 + 9y^2$

*Solution.*  Rearrange the terms so that like terms are in the same column. Note that

$$4y - 3x = -3x + 4y$$

Thus

$$\begin{array}{r} 4x + 5y \\ -3x + 4y \\ \hline x + 9y \end{array}$$

The correct choice is (C).                                                          ◀

**Example 6 ▶**    Find the sum of $2a^2 + 8a$ and $5 - 4a$.

(A)  $7a^2 + 4a$           (B)  $7a^2 - 4a$           (C)  $7a^2 + 4a + 5$

(D)  $2a^2 + 4a + 5$       (E)  $2a^2 - 4a + 5$

*Solution.*  Line up the terms as indicated.

$$\begin{array}{r} 2a^2 + 8a \\ - 4a + 5 \\ \hline 2a^2 + 4a + 5 \end{array}$$

The sum is $2a^2 + 4a + 5$. ◀

*Try the exercises for Topic E on page 158.*

## F. SUBTRACTING POLYNOMIALS

For any polynomial $P$, let $-P$ denote the polynomial obtained by changing the sign of every term of $P$. For example, let

$$P = 5x - 3y + 2$$

Then

$$-P = -5x + 3y - 2$$

Now add $P$ and $-P$.

$$
\begin{array}{ll}
P: & 5x - 3y + 2 \\
-P: & \underline{-5x + 3y - 2} \\
& 0
\end{array}
$$

Note that $0x + 0y + 0 = 0$.

In general, for any polynomial $P$,

$$\boxed{P + (-P) = (-P) + P = 0}$$

Recall that for numbers $a$ and $b$,

$$a - b = a + (-b)$$

This is also the way we subtract polynomials. For any polynomials $P$ and $Q$, define

$$P - Q = P + (-Q)$$

Thus, *to subtract the polynomial Q from P, add $-Q$.*

**Example 7** ▶ Subtract $4x + 7$ from $x^2 - 3$.

(A) $5x^2 + 4$       (B) $3x^2 + 4$       (C) $x^2 + 4x + 4$

(D) $-x^2 + 4x + 10$       (E) $x^2 - 4x - 10$

**Solution.** Here $P = x^2 - 3$ and $Q = 4x + 7$. (You are asked to subtract $Q$ from $P$.) Thus you want $P - Q$, or $P + (-Q)$.

$$-Q = -4x - 7$$

Line up like terms in the same column and add:

$$
\begin{array}{ll}
P: & x^2 \qquad - 3 \\
-Q: & \underline{\quad - 4x - 7} \\
& x^2 - 4x - 10
\end{array}
$$

The correct choice is (E). ◀

*Try the exercises for Topic F on page 159.*

**EXERCISES**

**A.**  *What is the coefficient of each term?*

1.  $20x^2$                    *Answer:*          2.  $-3ab$                  *Answer:*

3.  $x$                        *Answer:*          4.  $-t^3$                  *Answer:*

5.  $16$                       *Answer:*          6.  $(2x) \cdot (3y)$       *Answer:*

**B.**  *Fill in the terms.*

7.  $x^2 + 4x$ is the sum of the terms ☐ and ☐ .

8.  $a + 5$ is the sum of the terms ☐ and ☐ .

9.  $3b^2 - 5$ is the sum of the terms ☐ and ☐ .

10.  $2t^2 - 5t - 8$ is the sum of the terms ☐ , ☐ , and ☐ .

**C.**  *Which pairs are like terms?*

11.  $20x$ and $20y$      *Answer:*          12.  $3a$ and $7a$         *Answer:*

13.  $5b$ and $-5b$       *Answer:*          14.  $2c^2$ and $-c^2$     *Answer:*

15.  $xy$ and $2yx$       *Answer:*          16.  $xy^2$ and $2xy$      *Answer:*

17.  $3a^2$ and $2a^3$    *Answer:*          18.  $9$ and $-9$          *Answer:*

19.  $a^2b^2$ and $a \cdot b \cdot a \cdot b$   *Answer:*          20.  $c^3d$ and $cd^3$     *Answer:*

**D.**  *Add or subtract. If the given expression cannot be simplified, just write down this expression.*

21.  $8a + 4a$            *Answer:*          22.  $10b^2 + 10b^2$       *Answer:*

23.  $8x - 7x$            *Answer:*          24.  $6x - 5y$             *Answer:*

25.  $3x - 3x$            *Answer:*          26.  $4y^2 + (-4)y^2$      *Answer:*

27.  $3xy + 9xy$          *Answer:*          28.  $4xy^2 - 2x^2y$       *Answer:*

**E.**  *In Exercises 29-32, draw a circle around the correct letter.*

29.  Add $4x + 7y$ and $2x + 3y$.

(A)  $6x + 10y$          (B)  $2x + 4y$            (C)  $8x^2 + 21y^2$

(D)  $6x^2 + 10y^2$      (E)  $16xy$

30.  Add $5a + 3b$ and $2a - 6b$.

(A)  $7a + 3b$           (B)  $10a^2 - 18b^2$      (C)  $7a^2 - 3b^2$

(D)  $7a - 3b$           (E)  $10a - 3b$

31.  Add $4x + 9a$ and $3x - 9a$.

    (A) $x$                 (B) $7x + 18a$          (C) $7x - 18a$

    (D) $7x$            (E) $7x^2$

32.  Add $3x^3 + 5x$ and $6x^3 - x$.

    (A) $9x^6 + 4x^2$        (B) $9x^3 + 4x$         (C) $9x^3 - 4x$

    (D) $18x^3 - 5x$       (E) $18x^6 - 5x^2$

33.  Add $4a + 1$ and $-3a - 2$.  *Answer;*

34.  Add $3a + 5b$ and $10b - 2a$.  *Answer:*

35.  Add $9x - 2y$ and $4y - x$.  *Answer:*

36.  Find the sum of $4x^2 + 3$ and $6 - x^2$.  *Answer:*

37.  Find the sum of $3x + 2y$ and $2x - 3y$.  *Answer:*

38.  Find the sum of $9p - 3q$ and $6q - p$.  *Answer:*

39.  Find the sum of $4x^2 + 3x - 1$ and $2x^2 - x + 4$.  *Answer:*

40.  Find the sum of $6x + 3y - 2$ and $2x - xy + 5$.  *Answer:*

F.  *When several choices are given, draw a circle around the correct letter.*

  41.  Subtract $4x + 2y$ from $9x + 5y$.

      (A) $13x + 7y$        (B) $36x + 10y$       (C) $36x^2 + 10y^2$

      (D) $-5x - 3y$        (E) $5x + 3y$

  42.  Subtract $3a + 2b$ from $8a + 2b$.

      (A) $5a + 2b$         (B) $5a + 4b$         (C) $5a - 4b$

      (D) $5a$             (E) $11a + 4b$

  43.  Subtract $2p - 3q$ from $6p + 6q$.

      (A) $8p + 3q$         (B) $4p + 3q$         (C) $4p - 3q$

      (D) $4p + 9q$         (E) $-4p - 9q$

  44.  Subtract $6x - 3$ from $9x - 2$.

      (A) $3x - 5$          (B) $3x + 5$          (C) $3x + 1$

      (D) $3x - 1$          (E) $3x - 6$

45.  Subtract $2x - 5$ from $3x^2 + 2$.   *Answer:*

46.  Subtract $8a - 3b$ from $2a - 3$.   *Answer;*

47.  Subtract $5a - 2b$ from $a + 3c$.   *Answer:*

48.  Subtract $x - y$ from $x + y$.   *Answer:*

49.  Subtract $4x + 1$ from $x^2 - 3x$.   *Answer:*

50.  Subtract $5a - 1$ from $a^2 - 3$.   *Answer:*

51.  What is 5 less than $10x + 3$?   *Answer:*

   (A) $5x + 3$     (B) $10x + 8$     (C) $10x - 5$     (D) $10x - 2$     (E) $10x - 8$

52.  What is 3 less than $4a^2 - 6$?

   (A) $a^2 - 6$     (B) $a^2 - 9$     (C) $a^2 - 3$     (D) $4a^2 - 3$     (E) $4a^2 - 9$

53.  Let $x$ represent a number. Express the sum of two more than the number and three less than twice the number.

   (A) $x + 2$     (B) $3x - 2$     (C) $4x$     (D) $3x + 5$     (E) $3x - 1$

54.  Suppose that Barbara has $n$ nickels and Harriet has $d$ dimes. How much do they have together (in cents)?

   (A) $n + d$     (B) $5n + d$     (C) $n + 10d$     (D) $5n + 10d$     (E) $10d - 5n$

# 23.  Multiplying Polynomials

How are polynomials multiplied? To answer this, we first consider multiplying powers of a variable.

**A.** $a^m \cdot a^n$

Observe that

$$4^2 \underset{\lfloor 16 \rfloor}{\times} 4^1 = (4 \times 4) \times 4 = 4 \times 4 \times 4 = 4^3 \underset{\lfloor 64 \rfloor}{}$$

or

$$16 \times 4 = 64$$

Thus

$$4^2 \times 4^1 = 4^{2+1}$$

Now consider what happens when you multiply two powers of the same variable.

$$a^2 \cdot a^3 = (a \cdot a) \cdot (a \cdot a \cdot a) = a \cdot a \cdot a \cdot a \cdot a = a^5$$

Thus

$$a^2 \cdot a^3 = a^{2+3}$$

In general, for any positive integers $m$ and $n$ and for any number or variable $a$,

$$a^m \cdot a^n = a^{m+n}$$

**Example 1** ▶ Find $t^4 \cdot t^3$

*Solution.*

$$t^4 \cdot t^3 = t^{4+3} = t^7$$

by the above rule. Observe that if we multiply all the $t$'s, we also obtain

$$t^4 \cdot t^3 = (t \cdot t \cdot t \cdot t) \cdot (t \cdot t \cdot t)$$
$$= t \cdot t \cdot t \cdot t \cdot t \cdot t \cdot t$$
$$= t^7$$

Be sure you understand that

$$t^4 \cdot t^3 = t^{4+3} \qquad \text{and } not \ t^{4 \cdot 3}$$

because $t^{4 \cdot 3} = t^{12}$, which involves 12 uses of the factor $t$ instead of 7 such uses.    ◀

**Example 2** ▶ Find $y^6 \cdot y^4$.

    (A) $y^{24}$      (B) $y^{64}$      (C) $y^{10}$      (D) $y^{12}$      (E) $y^2$

*Solution.*

$$y^6 \cdot y^4 = y^{6+4} = y^{10}$$

The correct choice is (C).    ◀

**Example 3** ▶ Find $a \cdot a^4$.

*Solution.* Observe that $a = a^1$. Thus

$$a \cdot a^4 = a^1 \cdot a^4 = a^{1+4} = a^5$$   ◀

*Try the exercises for Topic A on page 164.*

## B. MULTIPLYING MONOMIALS

DEFINITION

A nonzero term is called a **monomial**.

*To multiply monomials*, use the associative and commutative laws in order to rearrange the factors as follows:

1. Group all coefficients at the beginning and multiply them.
2. Group powers of the same variable together. Multiply powers of each variable by the rule

$$a^m \cdot a^n = a^{m+n}$$

For example,

Group coefficients    Group powers of the
at the beginning.     same variable together.

$$(2xy^2)(3x^3y^4) = (2 \times 3)(x \cdot x^3)(y^2 \cdot y^4)$$
$$= 6x^{1+3}y^{2+4}$$
$$= 6x^4y^6$$

**Example 4** ▶   Multiply $5a^2b^3$ by $-2a^3b^5$.

(A) $3a^3b^5$    (B) $-10a^6b^{15}$    (C) $3a^6b^{15}$    (D) $-10a^5b^8$    (E) $3a^5b^8$

*Solution.*

$$(5a^2b^3)(-2a^3b^5) = (5 \times (-2))(a^2 \cdot a^3)(b^3 \cdot b^5)$$
$$= -10a^{2+3}b^{3+5}$$
$$= -10a^5b^8$$

The correct choice is (D).   ◀

*Try the exercises for Topic B on page 165.*

## C. MULTIPLYING A BINOMIAL BY A MONOMIAL

DEFINITION

The sum (or difference) of two *unlike* monomials is called a **binomial**.

Thus

$$5a + 3b, \qquad 4x - 3, \qquad \text{and} \qquad y^2 + 4y$$

are each binomials, whereas

$$7t + 2t, \qquad \text{which equals } 9t$$

is a monomial.

Recall that

$$ax + bx = (a + b)x$$

This rule enabled us to add like terms. More generally, let $a, b$, and $c$ be numbers or terms. Then the **Distributive Laws** assert that

$$a(b + c) = ab + ac$$
$$a(b - c) = ab - ac$$
$$(a + b)c = ac + bc$$
$$(a - b)c = ac - bc$$

*To multiply a binomial by a monomial* (in either order), use the distributive laws. For example, use

$$a(b + c) = ab + ac$$

to obtain

$$4x(x + y) = 4x \cdot x + 4x \cdot y = 4x^2 + 4xy$$

**Example 5 ▶**  Multiply.        $4p(p^2q - pq^3)$

**Solution.**  Use the second distributive law:
$$4p(p^2q - pq^3) = 4p \cdot p^2q - 4p \cdot pq^3$$
$$= 4p^3q - 4p^2q^3 \qquad ◀$$

**Example 6 ▶**  Find the product of $2x - 3x^2y$ and $5x^4y$.

**Solution.**  Use the fourth distributive law.
$$(2x - 3x^2y)(5x^4y) = 2x \cdot 5x^4y - 3x^2y \cdot 5x^4y$$
$$= (2 \times 5)(x \cdot x^4)y - (3 \times 5)(x^2 \cdot x^4)(y \cdot y)$$
$$= 10x^5y - 15x^6y^2 \qquad ◀$$

*Try the exercises for Topic C on page 165.*

## D. ADDITION, SUBTRACTION, AND MULTIPLICATION

In some examples you must multiply and add, or multiply and subtract. Recall that you multiply before you add or subtract. Thus

$$x^2 + x(x + y) = x^2 + x \cdot x + x \cdot y \qquad \text{First multiply } x \text{ by } x + y.$$
$$\text{Now add } x^2 \text{ and } x^2.$$
$$= x^2 + x^2 + xy$$
$$= 2x^2 + xy$$

**Example 7 ▶**    Simplify.    $4ab^2 - 2ab(3b - 2a)$

(A) $4ab^2 - 2ab + 3b - 2a$    (B) $4ab^2 - 6ab^2 - 2a$

(C) $4ab^2 - 6ab^2 + 4a^2b$    (D) $-2ab^2 - 2a$    (E) $-2ab^2 + 4a^2b$

**Solution.** First multiply $2ab$ by $3b - 2a$. Then subtract the result from $4ab^2$.

$$4ab^2 - 2ab(3b - 2a) = 4ab^2 - (2ab \cdot 3b - 2ab \cdot 2a)$$
$$= 4ab^2 - \big((2 \times 3)a(b \cdot b) - (2 \times 2)(a \cdot a)b\big)$$
$$= 4ab^2 - (6ab^2 - 4a^2b)$$

$$\text{minus} \times \text{minus}$$
$$= 4ab^2 - 6ab^2 + 4a^2b$$
$$\underline{\phantom{xx}-2ab^2\phantom{xx}}$$

$$= -2ab^2 + 4a^2b$$

The correct choice is (E).    ◀

**Example 8 ▶**    Simplify.    $4pq - 2p(pq^2 + p^2q)$

**Solution.** First multiply $2p$ by $pq^2 + p^2q$. Then subtract the result from $4pq$.

$$4pq - 2p(pq^2 + p^2q) = 4pq - (2p \cdot pq^2 + 2p \cdot p^2q)$$
$$= 4pq - (2p^2q^2 + 2p^3q)$$
$$= 4pq - 2p^2q^2 - 2p^3q$$

There are three (mutually) unlike terms. Thus the result cannot be simplified any further.    ◀

*Try the exercises for Topic D on page 166.*

## EXERCISES

**A.** *Find each product. In Exercises 1-4, draw a circle around the correct letter.*

1.  $a^3 \cdot a^4$

    (A) $a^5$    (B) $a^7$    (C) $a^{12}$    (D) $a^3 + a^4$    (E) $a^{34}$

2.  $b^5 \cdot b^2$

    (A) $b^5$    (B) $b^{10}$    (C) $b^9$    (D) $b^7$    (E) $b^{25}$

3.  $x \cdot x^3$

    (A) $(x^2)^3$    (B) $x^2$    (C) $x^3$    (D) $x^4$    (E) $x^6$

4.  $y^7 \cdot y^3$

    (A) $y^{21}$    (B) $y^{14}$    (C) $y^{10}$    (D) $y^7$    (E) $y^3$

5. $m^4 \cdot m^4 =$        6. $n^9 \cdot n =$        7. $t^6 \cdot t^5 =$        8. $z^{10} \cdot z^{10} =$

**B.** *Find each product. In Exercises 9-12, draw a circle around the correct letter.*

9.  The product of $4a^2$ and $ab$ is

    (A) $4a^2b$        (B) $4a^3b$        (C) $4a^4b$        (D) $4a^4b^4$        (E) $a^3b^4$

10. The product of $3c^2d$ and $2cd$ is

    (A) $6c^2d^2$        (B) $6cd$        (C) $6c^3d$        (D) $6c^3d^2$        (E) $6c^2d^3$

11. What is the product of $9xy^2$ and $3x^4$?

    (A) $27x^5y^2$        (B) $27x^6y^2$        (C) $12xy^6$        (D) $27x^4y^2$        (E) $27(xy)^5$

12. What is the product of $4p^2q$ and $5pq^2$?

    (A) $20p^3q^3$        (B) $9p^3q^3$        (C) $20p^2q^3$        (D) $20p^3q^2$        (E) $20p^4q^3$

13. Find the product of $5m^4n$ and $-2mn$.    *Answer:*

14. Find the product of $-x^3y^2$ and $-x^2y^3$.    *Answer:*

15. The product of $7p^2q^5$ and $3p^4q$ is    _____

16. The product of $-5a^6b$ and $-6ab^3$ is    _____

17. The product of $2ab^2c$ and $3ab$ is    _____

18. The product of $2x^2yz^3$ and $-4xyz$ is    _____

**C.** *Find each product. In Exercises 19-24, draw a circle around the correct letter.*

19. Find the product of $5a^2bc^4$ and $-3a^2c^4$.

    (A) $2a^2bc^4$            (B) $2a^2c^4$            (C) $-15a^2bc^4$

    (D) $-15a^4bc^8$            (E) $-15a^4bc^{16}$

20. Find the product of $9r^2st^6$ and $-3r^4t^4$.

    (A) $6r^6st^{10}$            (B) $-27r^8st^{24}$            (C) $-27r^6st^{24}$

    (D) $-27r^8st^{10}$            (E) $-27r^6st^{10}$

21. The product of $-4x^3y^2z^4$ and $-8x^2yz^6$ is

    (A) $-12x^5y^3z^{10}$            (B) $-32x^5y^3z^{24}$            (C) $32x^5y^3z^{24}$

    (D) $12x^5y^3z^{10}$            (E) $32x^5y^3z^{10}$

22.  The product of $-a^4b^5c^3$ and $-2ab^6c$ is

(A)  $-3a^5b^{11}c^4$

(B)  $-2a^5b^{11}c^4$

(C)  $2a^5b^{11}c^4$

(D)  $2a^4b^{30}c^3$

(E)  $2a^5b^{30}c^4$

23.  $a(a + 3) =$

(A)  $a^2 + 3$

(B)  $a^2 + 3a$

(C)  $2a + 3$

(D)  $4a^2$

(E)  $4a$

24.  $2b(b - 5) =$

(A)  $2b - 10$

(B)  $2b^2 - 10$

(C)  $2b^2 - 5$

(D)  $2b^2 - 10b$

(E)  $2b^2 - 5b$

25.  $(c^2 + 4)3c =$

26.  $(x^2 - 2x)5x =$

27.  Multiply $y^2$ by $4y - 3$.    *Answer:*

28.  Multiply $6z^2 - z$ by $-3z$.    *Answer:*

29.  Find the product of $4t^2$ and $2t - 3$.    *Answer:*

30.  Find the product of $2pq$ and $p - q$.    *Answer:*

31.  $3x^2y(2x - 1) =$

32.  $9ab^2(ab + 1) =$

33.  $-4xy^3(x^2 - xy) =$

34.  $10a^2bc(a^2c - bc^2) =$

D.  *Simplify each polynomial. In Exercises 35-40, draw a circle around the correct letter.*

35.  $2x + x(x + 1) =$

(A)  $2x + x^2 + 1$

(B)  $x^2 + 3x$

(C)  $x^2 + 3x + 1$

(D)  $x^2 + x$

(E)  $x^2 - x$

36.  $4y + y^2(y + 1) =$

(A)  $4y + y^3 + y^2$

(B)  $4y + y^3 + 1$

(C)  $y^3 + 5y$

(D)  $y^3 + 5y^2$

(E)  $y^3 + 5y + 1$

37.  $6x^2 - 2x(x + 1) =$

    (A) $4x^2 + 1$         (B) $4x^2 - 2x$         (C) $4x^2 - 2x + 1$

    (D) $4x^2 + 2x$        (E) $6x^2 - 2x - 1$

38.  $5a^2 - 3a(a - 1) =$

    (A) $5a^2 - 3a - 1$      (B) $5a^2 - 3a + 1$      (C) $5a^2 + 3a$

    (D) $2a^2 - 3a$        (E) $2a^2 + 3a$

39.  $x^2y - x(xy - 1) =$

    (A) $x^2y - xy^2 - 1$     (B) $x^2y - xy^2 + 1$     (C) $x^2y + 1$

    (D) $2x^2y + x$        (E) $x$

40.  $p^2q - 4p(p^2 - pq) =$

    (A) $p^2q - 4p^3 - pq$     (B) $p^2q - 4p^3 - 4pq$     (C) $5p^2q - 4p^3$

    (D) $-3p^2q - 4p^3$     (E) $-4p^3$

41.  $6x^2y^2 - 4x(x^2y - xy^2) =$

42.  $8a^2b - 5ab(2a - 1) =$

43.  $6xy - 2y(x^2 - x) =$

44.  $4a^2b + 3a(ab - 1) =$

45.  $10x^2y^2 - 3xy(x - 2xy) =$

46.  $12a^2b^3 - 2ab^2(ab - 3) =$

# 24.  Powers of Monomials

It is important to know how to raise a monomial to a power.

## A. POWERS OF POWERS

Recall that in the expression

$$2^3 = 8$$

2 is the *base*, 3 is the *exponent* of 2, and $2^3$, or 8, is the 3rd *power* of 2. What is $(x^3)^2$? Observe that

$$(x^3)^2 = x^3 \cdot x^3$$

$$= (x \cdot x \cdot x) \cdot (x \cdot x \cdot x)$$

$$= \underbrace{x \cdot x \cdot x \cdot x \cdot x \cdot x}_{\text{6 factors}}$$

$$= x^6$$

Thus

$$(x^3)^2 = x^6 = x^{3 \times 2}$$

Similarly,

$$(a^2)^4 = \underbrace{a^2 \cdot a^2 \cdot a^2 \cdot a^2}_{\text{4 such factors}}$$

$$= (a \cdot a) \cdot (a \cdot a) \cdot (a \cdot a) \cdot (a \cdot a)$$

$$= \underbrace{a \cdot a \cdot a \cdot a \cdot a \cdot a \cdot a \cdot a}_{\text{8 factors}}$$

$$= a^8$$

Thus

$$(a^2)^4 = a^8 = a^{2 \times 4}$$

> To raise a power to a power, write the base, and multiply the exponents. In symbols,
>
> $$(a^m)^n = a^{mn}$$

In particular, if $m = 2$ and $n = 4$, then, as shown above,

$$(a^2)^4 = a^{2 \times 4}$$

**Example 1 ▶**   Find $(b^4)^3$.

(A) $b^7$          (B) $b^8$          (C) $b^{12}$          (D) $b^{16}$          (E) $b^{64}$

***Solution.***   Write the base $b$, and multiply the exponents 4 and 3.

$$(b^4)^3 = b^{4 \times 3} = b^{12}$$

The correct choice is (C).          ◀

**Example 2 ▶**   Find $(c^5)^4$.

***Solution.***   Write the base $c$, and multiply the exponents 5 and 4.

$$(c^5)^4 = c^{5 \times 4} = c^{20}$$          ◀

*Try the exercises for Topic A on page 171.*

## B. POWERS OF PRODUCTS

By the commutative and associative laws, we can rearrange the factors when multiplying. Thus

$$(ab)^3 = ab \cdot ab \cdot ab$$
$$= (a \cdot a \cdot a)(b \cdot b \cdot b)$$
$$= a^3 b^3$$

In general,

> A power of a product is the product of the powers. In symbols,
>
> $$(ab)^n = a^n b^n$$

Similarly,

$$(x^4 y^3)^2 = (x^4 y^3) \cdot (x^4 y^3)$$
$$= (x \cdot x \cdot x \cdot x \cdot y \cdot y \cdot y)(x \cdot x \cdot x \cdot x \cdot y \cdot y \cdot y)$$
$$= (x \cdot x \cdot x \cdot x \cdot x \cdot x \cdot x \cdot x)(y \cdot y \cdot y \cdot y \cdot y \cdot y)$$
$$= x^8 y^6$$

In other words,

$$(x^4y^3)^2 = x^8y^6 = x^{4\times 2}y^{3\times 2}$$

In general:

$$(a^mb^n)^r = a^{m\times r}b^{n\times r}$$

In the preceding illustration,

$$a = x, \quad b = y, \quad m = 4, \quad n = 3, \quad r = 2$$

**Example 3** ▶  $(st^3)^3 =$

(A) $st^6$        (B) $st^9$        (C) $st^{27}$        (D) $s^3t^{27}$        (E) $s^3t^9$

*Solution.*

$$(st^3)^3 = (s^1t^3)^3$$
$$= s^{1\times 3}t^{3\times 3}$$
$$= s^3t^9$$

The correct choice is (E).                                                                            ◀

The preceding rule applies to products of several different factors. For example,

$$(a^mb^nc^p)^r = a^{m\times r}b^{n\times r}c^{p\times r}$$

**Example 4** ▶  $(2xy^2)^3 =$

(A) $2x^3y^5$      (B) $2x^3y^6$      (C) $8x^3y^8$      (D) $8x^3y^6$      (E) $8x^3y^5$

*Solution.*

$$(2xy^2)^3 = 2^3x^3(y^2)^3$$
$$= 8x^3y^{2\times 3}$$
$$= 8x^3y^6$$

The correct choice is (D).                                                                            ◀

*Try the exercises for Topic B on page 172.*

## C. MULTIPLYING AND RAISING TO A POWER

Often you must multiply powers and raise a power to a power in the same example. Recall that

$$a^m \cdot a^n = a^{m+n} \quad \text{whereas} \quad (a^m)^n = a^{mn}$$

**Example 5** ▶  Simplify.   $a^2(10ab^4)^3$

(A) $10a^3b^{12}$                (B) $1000a^3b^{12}$                (C) $1000a^3b^7$

(D) $1000a^3b^{64}$                (E) $1000a^5b^{12}$

**Solution.** First raise to a power; then multiply.

$$a^2 (10ab^4)^3 = a^2 \cdot 10^3 a^3 b^{4 \times 3}$$
$$= 1000 a^{2+3} b^{12}$$
$$= 1000 a^5 b^{12}$$

Note that

$$a^2 \cdot a^3 = a^{2+3} = a^5$$

Here you *multiply* two powers of $a$. Write the base $a$, and *add* the exponents. On the other hand,

$$(b^4)^3 = b^{4 \times 3} = b^{12}$$

Here you raise a power to a power. Write the base $b$, and *multiply* the exponents. The correct choice is (E).      ◄

*Try the exercises for Topic C on page 172.*

### EXERCISES

**A.** *Find each power.*

1. $(s^4)^2 =$        2. $(t^3)^4 =$        3. $(u^6)^2 =$

4. $(x^9)^2 =$        5. $(y^2)^5 =$        6. $(z^4)^6 =$

*Draw a circle around the correct letter.*

7. $(a^5)^3 =$

    (A) $a^2$        (B) $a^8$        (C) $a^{10}$        (D) $a^{15}$        (E) $a^{125}$

8. $(b^{10})^2 =$

    (A) $b^{10}$        (B) $b^{11}$        (C) $b^{20}$        (D) $b^{100}$        (E) $b^{200}$

9. $(c^7)^3 =$

    (A) $c^{21}$        (B) $c^{14}$        (C) $c^7$        (D) $c^4$        (E) $c$

10. $(x^3)^8 =$

    (A) $x^3$        (B) $x^8$        (C) $x^9$        (D) $x^{27}$        (E) $x^{24}$

11. $(y^6)^5 =$

    (A) $y^6$        (B) $y^{11}$        (C) $y^{12}$        (D) $y^{18}$        (E) $y^{30}$

12.  $(z^{10})^5 =$

   (A) $z^{100}$         (B) $z^{50}$         (C) $z^{15}$         (D) $z^5$         (E) $z^{10,000}$

**B.** *In Exercises 13-18, draw a circle around the correct letter.*

13.  $(ab^2)^2 =$

   (A) $a^2b^4$         (B) $ab^4$         (C) $ab^8$         (D) $a^2b^8$         (E) $a^2b^6$

14.  $(c^2d^3)^2 =$

   (A) $c^2d^5$         (B) $c^2d^6$         (C) $c^2d^9$         (D) $c^4d^5$         (E) $c^4d^6$

15.  $(p^3q)^4 =$

   (A) $p^3q^4$         (B) $p^7q^4$         (C) $p^{12}q^4$         (D) $p^{81}q^4$         (E) $p^{81}q$

16.  $(s^2t^5)^3 =$

   (A) $s^6t^{15}$         (B) $s^8t^{15}$         (C) $s^8t^{125}$         (D) $s^2t^{15}$         (E) $s^2t^8$

17.  $(2a^3)^2 =$

   (A) $2a^5$         (B) $2a^6$         (C) $2a^8$         (D) $4a^5$         (E) $4a^6$

18.  $(3x^4)^2 =$

   (A) $3x^8$         (B) $9x^8$         (C) $3x^6$         (D) $9x^6$         (E) $3x^{16}$

19.  $(5y^3)^2 =$                          20.  $(-2t^4)^2 =$

21.  $(10u^5)^3 =$                          22.  $(-v^6)^3 =$

23.  $(3x^2)^3 =$                          24.  $(4y^4)^3 =$

25.  $(2xy^3)^2 =$                          26.  $(6a^2b^2)^2 =$

27.  $(5pq^4)^2 =$                          28.  $(3a^3b^2)^3 =$

29.  $(-10x^2y^4)^2 =$                          30.  $(5a^3b^4)^3 =$

**C.** *Simplify. In Exercises 31-34, draw a circle around the correct letter.*

31.  $a(a^2b)^2 =$

   (A) $a^4b^2$         (B) $a^3b^2$         (C) $a^2b^2$         (D) $a^5b^2$         (E) $a^6b^2$

32. $b(4b^2)^2 =$

    (A) $4b^4$      (B) $4b^5$      (C) $8b^5$      (D) $16b^5$      (E) $16b^6$

33. $x^2(10x^3)^2 =$

    (A) $10x^6$      (B) $10x^7$      (C) $10x^8$      (D) $100x^7$      (E) $100x^8$

34. $a^3(2ab^2)^3 =$

    (A) $2a^4b^5$      (B) $2a^4b^6$      (C) $2a^6b^6$      (D) $8a^4b^6$      (E) $8a^6b^6$

35. $x^4(5x^2y)^2 =$                36. $p^2(3p^2q)^2 =$

37. $x^5(3xy^2)^3 =$                38. $2x(2x^2y^4)^2 =$

39. $ab(2a^2b^3)^2 =$               40. $x^2y^3(3x^3y^2)^3 =$

# 25.  Dividing Polynomials

We will consider how to divide monomials and how to divide a binomial by a monomial.

## A.  DIVIDING MONOMIALS

When dividing monomials, express division by means of a fraction, and divide by all factors common to the numerator and denominator.

**Example 1** ▶   Divide $6x$ by 3.

*Solution.* Express this as the fraction

$$\frac{6x}{3}$$

Divide 6 by 3.

$$\frac{\overset{2}{\cancel{6}}x}{\underset{1}{\cancel{3}}} = 2x$$

The quotient is $2x$.    ◀

**Example 2** ▶    Find $\dfrac{7a^2}{a}$.

*Solution.*

$$\frac{7a^2}{a} = \frac{7a \cdot \overset{1}{\cancel{a}}}{\underset{1}{\cancel{a}}} = 7a$$

The quotient is $7a$.    ◀

**Example 3** ▶    Find $\dfrac{8x^2}{4x}$.

*Solution.* Divide 8 by 4 and $x^2$ by $x$.

$$\frac{8x^2}{4x} = \frac{\overset{2 \cdot 1}{\cancel{8}\cancel{x} \cdot x}}{\underset{1 \cdot 1}{\cancel{4}\cancel{x}}} = 2x$$

The quotient is $2x$.    ◀

*Try the exercises for Topic A on page 176.*

## B. DIVIDING A BINOMIAL BY A MONOMIAL

To add two fractions with the same denominator, we write the denominator and add the numerators. Thus

$$\frac{2}{7} + \frac{4}{7} = \frac{2+4}{7} = \frac{6}{7}$$

In general,

$$\frac{a}{c} + \frac{b}{c} = \frac{a+b}{c}$$

and

$$\frac{a}{c} - \frac{b}{c} = \frac{a-b}{c}$$

A binomial, such as $ax + b$, can be divided by $c$ by breaking up the fraction $\dfrac{ax + b}{c}$ into two fractions:

$$\frac{ax + b}{c} = \frac{ax}{c} + \frac{b}{c}$$

Thus

$$\frac{2x + 4}{2} = \frac{\overset{1}{\cancel{2}x}}{\underset{1}{\cancel{2}}} + \frac{\overset{2}{\cancel{4}}}{\underset{1}{\cancel{2}}} = x + 2$$

**Example 4 ▶**  $\dfrac{4x + 12}{4} =$

(A) $x + 12$    (B) $x + 3$    (C) $4x + 3$    (D) $x + 8$    (E) $4x + 48$

*Solution.*

$$\frac{4x + 12}{4} = \frac{\overset{1}{\cancel{4}x}}{\underset{1}{\cancel{4}}} + \frac{\overset{3}{\cancel{12}}}{\underset{1}{\cancel{4}}} = x + 3$$

The quotient is $x + 3$. The correct choice is **(B)**.    ◀

**Example 5 ▶**  $\dfrac{5a^2 + 7a}{a} =$

*Solution.*

$$\frac{5a^2 + 7a}{a} = \frac{5a^2}{a} + \frac{7a}{a}$$

$$= \frac{5a \cdot \overset{1}{\cancel{a}}}{\underset{1}{\cancel{a}}} + \frac{7\overset{1}{\cancel{a}}}{\underset{1}{\cancel{a}}}$$

$$= 5a + 7$$

The quotient is $5a + 7$.    ◀

**Example 6 ▶**  $\dfrac{10y^2 - 15y}{5y} =$

(A) $2y$    (B) $2y - 3y$    (C) $-y$    (D) $2y - 3$    (E) $2y - 5$

*Solution.*

$$\frac{10y^2 - 15y}{5y} = \frac{10y^2}{5y} - \frac{15y}{5y}$$

$$= \frac{\overset{2 \cdot 1}{\cancel{10y} \cdot y}}{\underset{1 \cdot 1}{\cancel{5y}}} - \frac{\overset{3 \cdot 1}{\cancel{15y}}}{\underset{1 \cdot 1}{\cancel{5y}}}$$

$$= 2y - 3$$

The quotient is $2y - 3$. The correct choice is (D).   ◀

*Try the exercises for Topic B on this page.*

## EXERCISES

**A.** *Divide.*

1. $\dfrac{4a}{4}$   *Answer:*          2. $\dfrac{9b}{9}$   *Answer:*

3. $\dfrac{15c}{5}$   *Answer:*          4. $\dfrac{10x}{2}$   *Answer:*

5. $\dfrac{9x^2}{3}$   *Answer:*          6. $\dfrac{24y^2}{8}$   *Answer:*

7. $\dfrac{-20m^2}{10}$   *Answer:*          8. $\dfrac{42n^4}{-6}$   *Answer:*

9. $\dfrac{x^2}{x}$   *Answer:*          10. $\dfrac{y^3}{y}$   *Answer:*

11. $\dfrac{2s^4}{s}$   *Answer:*          12. $\dfrac{5t^3}{t}$   *Answer:*

13. $\dfrac{-6x^4}{x}$   *Answer:*          14. $\dfrac{7a^8}{a}$   *Answer:*

15. $\dfrac{2a}{2a}$   *Answer:*          16. $\dfrac{-5b}{5b}$   *Answer:*

17. $\dfrac{4a}{2a}$   *Answer:*          18. $\dfrac{16b}{4b}$   *Answer:*

19. $\dfrac{10c^2}{10c}$   *Answer:*          20. $\dfrac{20d^2}{10d}$   *Answer:*

21. $\dfrac{15x^3}{3x}$   *Answer:*          22. $\dfrac{28y^4}{7y}$   *Answer:*

23. $\dfrac{-32s^5}{4s}$   *Answer:*          24. $\dfrac{72t^8}{12t}$   *Answer:*

**B.** *Divide. When several choices are given, draw a circle around the correct letter.*

25. $\dfrac{2a + 2}{2} =$

   (A) $a$          (B) 1          (C) $a + 2$          (D) $a + 1$          (E) 2

26. $\dfrac{4x - 4}{4} =$

   (A) $x - 1$          (B) $x$          (C) 1          (D) $-1$          (E) $x - 4$

27. $\dfrac{10t + 10}{5} =$

(A) 2      (B) $2t$      (C) $2t + 1$      (D) $2t + 2$      (E) $2t + 10$

28. $\dfrac{10y + 20}{10} =$

(A) $y + 2$      (B) $y + 1$      (C) $y + 20$      (D) $y + 10$      (E) $y$

29. $\dfrac{5z - 15}{5} =$

30. $\dfrac{12a + 18}{6} =$

31. $\dfrac{9b - 15}{3} =$

32. $\dfrac{12b^2 - 24}{6} =$

33. $\dfrac{a^2 + a}{a} =$

(A) $a$      (B) $a^2 + 1$      (C) $a + 1$      (D) $2a$      (E) $a - 1$

34. $\dfrac{b^2 - b}{b} =$

(A) $b + 1$      (B) $b - 1$      (C) $b^2 + 1$      (D) $b^2 - 1$      (E) 0

35. $\dfrac{c^3 + c}{c} =$

36. $\dfrac{d^3 - d^2}{d} =$

37. $\dfrac{2x^2 + 5x}{x} =$

38. $\dfrac{4y^2 + 6y}{y} =$

39. $\dfrac{2x^2 + 2x}{2x} =$

(A) 1      (B) $x$      (C) $x + 1$      (D) $x + 2$      (E) $x^2 + 1$

40. $\dfrac{5y^2 + 10y}{5y} =$

(A) $y + 10$      (B) $y + 2$      (C) $y + 1$      (D) $y$      (E) $y^2 + 2$

41. $\dfrac{6x^2 + 9x}{3x} =$

(A) $2x + 1$      (B) $2x + 3$      (C) $2x^2 + 3$      (D) $2x^2 + 3x$      (E) $2x$

42. $\dfrac{16y^2 + 8y}{8y} =$

(A) $2y$      (B) 1      (C) $2y^2 + 1$      (D) $2y + 1$      (E) $2y - 1$

43. $\dfrac{10x^2 - 15x}{5x} =$

44. $\dfrac{6a^2 - 18a}{3a} =$

45. $\dfrac{9b^3 - 12b}{3b} =$

46. $\dfrac{10c^3 + 20c}{5c} =$

47. $\dfrac{12x^4 + 16x^2}{4x} =$

48. $\dfrac{9y^5 - 15y^3}{3y} =$

# 26.   Common Factors

Factoring is an important way of simplifying a polynomial. We will consider a basic method known as **common factoring**.

## A.  GREATEST COMMON DIVISORS

Recall that when two or more numbers are multiplied, each is called a *factor*, or a *divisor*, of the product. For example,

$$10 = 2 \times 5$$

so that 2 and 5 are each factors (or divisors) of 10. Note that 5 is also a factor of 15 because

$$15 = 3 \times 5$$

We will say that 5 is a *common factor* of 10 and 15, according to the following definition:

DEFINITION

> Let $m$ and $n$ be any integers—positive, negative, or 0. Then the positive integer $a$ is called a **common factor** (or a **common divisor**) **of** $m$ **and** $n$ if $a$ is a factor of *both* $m$ and $n$. The largest common factor of $m$ and $n$ is called the **greatest common divisor of** $m$ **and** $n$.

Write

$gcd\ (m, n)$      for      the greatest common divisor of $m$ and $n$

**Example 1 ▶**  Find *gcd* (6, 9).

*Solution.*

1. The positive factors of 6 are 1, 2, 3, and 6.
2. The positive factors of 9 are 1, 3, and 9.
3. The *common* factors of 6 and 9 are 1 and 3.
4. The larger of these common factors is 3. Thus

$$gcd\ (6, 9) = 3$$   ◀

**Example 2 ▶**  Find *gcd* (8, 12).

*Solution.*

1. The positive factors of 8 are 1, 2, 4, and 8.
2. The positive factors of 12 are 1, 2, 3, 4, 6, and 12.
3. The *common* factors of 8 and 12 are 1, 2, and 4.
4. The largest of the common factors is 4. Thus

$$gcd\ (8, 12) = 4$$   ◀

**Example 3 ▶**  Find the greatest common divisor of the *coefficients* of $3x - 6$.

*Solution.*  The coefficients of $3x - 6$ are 3 and -6. Thus we want

$$gcd\ (3, -6)$$

1. The *positive* factors of 3 are 1 and 3.
2. The *positive* factors of -6 are 1, 2, 3, and 6.
3. The common factors of 3 and -6 are 1 and 3.
4. The largest of the common factors is 3. Thus

$$gcd\ (3, -6) = 3$$

and 3 is the greatest common divisor of the coefficients of $3x - 6$.   ◀

*Try the exercises for Topic A on page 182.*

## B. FACTORS OF A BINOMIAL

We want to say that 4 and $x$ are each *factors of the polynomial* $4x$.

DEFINITION

Let $P$ and $Q$ be polynomials, where $Q \neq 0$. Then $Q$ is called a **factor of $P$** if

$$P = Q \cdot R$$

for some polynomial $R$.

**Example 4 ▶**  Show that 2 is a factor of $2x + 8$.

*Solution.*  First note that 2 is the *gcd* of the coefficients of $2x + 8$. Also,

$$\frac{2x + 8}{2} = \frac{2x}{2} + \frac{8}{2} = x + 4$$

Thus

$$2x + 8 = 2(x + 4)$$
$$\underbrace{\quad}_{P} \quad \underset{Q}{\cdot} \underbrace{\quad}_{R}$$

and, according to the definition, 2 is a factor of $2x + 8$.    ◀

**Example 5** ▶ . Show that $a$ is a factor of

**Solution**.

$$\frac{3a^2 - 5a}{a} = \frac{3a^2}{a} - \frac{5a}{a}$$

$$= \frac{3a \cdot \overset{1}{\cancel{a}}}{\underset{1}{\cancel{a}}} - \frac{5\overset{1}{\cancel{a}}}{\underset{1}{\cancel{a}}}$$

$$= 3a - 5$$

Thus

$$3a^2 - 5a = a(3a - 5)$$
$$\underbrace{\quad}_{P} \quad \underset{Q}{\cdot} \underbrace{\quad}_{R}$$

and, according to the definition, $a$ is a factor of $3a^2 - 5a$.    ◀

*Try the exercises for Topic B on page 183.*

## C.  ISOLATING THE COMMON FACTOR

Clearly 5 and $x$ are each factors of

$$5x^2 + 10x$$

because

$$5x^2 + 10x = 5(x^2 + 2x)$$

and

$$5x^2 + 10x = x(5x + 10)$$

We will speak of $5x$, the product of 5 and $x$, as the *common factor* of $5x^2 + 10x$, according to the following definition:

**DEFINITION**

> Let $P$ be a binomial. Then the **common factor of** $P$ is defined as the *product* of
>
> 1. the *gcd* of the coefficients of $P$
>
>    and
>
> 2. the *smallest* power of each variable that occurs in *both* terms of $P$.

**Example 6** ▶    Find the common factor of $4x^2 + 6ax$.

*Solution.*

1. The *gcd* of the coefficients of $4x^2 + 6ax$ is 2, that is,

   $$gcd\ (4, 6) = 2$$

2. The only variable that occurs in *both* terms of $4x^2 + 6ax$ is $x$. The smallest power of $x$ that occurs is $x$ (in the second term $6ax$).

3. The product of 2 and $x$ is $2x$. Thus $2x$ is the common factor of $4x^2 + 6ax$.    ◀

DEFINITION

> To **isolate the common factor** $a$ of a binomial of the form $ab + ac$ or $ab - ac$, write
>
> $$ab + ac = a(b + c)$$
>
> or
>
> $$ab - ac = a(b - c)$$

For example,

$$4x + 4 = 4x + 4 \cdot 1 = 4(x + 1)$$
$$ab + a \cdot c = a(b + c)$$

$$4x + 8 = 4x + 4 \cdot 2 = 4(x + 2)$$
$$ab + a \cdot c = a(b + c)$$

$$4x - 12 = 4x - 4 \cdot 3 = 4(x - 3)$$
$$ab - a \cdot c = a(b - c)$$

$$4x^2 + 16x = 4x \cdot x + 4x \cdot 4 = 4x(x + 4)$$
$$\lfloor a \rfloor \cdot b + \lfloor a \rfloor \cdot c = \lfloor a \rfloor(b + c)$$

In order to see how to write a binomial in the form $ab + ac$ or $ab - ac$, we must first find the common factor, $a$.

**Example 7** ▶    Factor completely.    $5t^2 - 5t$

(A)  $5(t^2 - 1)$          (B)  $5t(t - 1)$          (C)  $5t^2(t - 1)$

(D)  $5t(t + 1)$          (E)  $5(t^2 - 5t)$

*Solution.*  First find the common factor of $5t^2 - 5t$.

1. The *gcd* of the coefficients of $5t^2 - 5t$ is 5.

2. The only variable that occurs in *both* terms is $t$, and the smallest power of $t$ that occurs is $t$ (or $t^1$).

3. The product of 5 and $t$ is $5t$, and thus $5t$ is the common factor of $5t^2 - 5t$.

   Now divide $5t^2 - 5t$ by the common factor, $5t$.

   $$\frac{5t^2 - 5t}{5t} = \frac{5t^2}{5t} - \frac{5t}{5t} = t - 1$$

Thus

$$5t^2 - 5t = 5t \cdot t - 5t \cdot 1 = 5t(t - 1)$$
$$a \cdot b - a \cdot c = a(b - c)$$

The correct choice is (B).    ◄

**Example 8** ▶ Factor.    $9ab^2 - 12ab$

> *Solution.* Even though the word *completely* is not used in the directions, you are expected to factor this binomial *completely*.
>
> 1. The *gcd* of the coefficients of $9ab^2 - 12ab$ is 3.
>
>    $gcd(9, 12) = 3$
>
> 2. The variables that occur in both terms are $a$ and $b$. The *smallest* powers of these variables are $a$ and $b$ (in the second term, $-12ab$).
>
> 3. The product of $3, a,$ and $b$, is $3ab$. Thus $3ab$ is the common factor.
>
> 4.    $$\frac{9ab^2 - 12ab}{3ab} = \frac{9ab^2}{3ab} - \frac{12ab}{3ab}$$
>
>    $$= 3b - 4$$

so that

$$9ab^2 - 12ab = 3ab(3b - 4)$$    ◄

*Try the exercises for Topic C on page 183.*

**EXERCISES**

**A.** *Find the greatest common divisor of each pair of integers.*

1. $gcd(4, 8)$    *Answer:*          2. $gcd(5, 10)$    *Answer:*

3. $gcd(6, 12)$    *Answer:*         4. $gcd(7, -14)$    *Answer:*

5. $gcd(4, 6)$    *Answer:*          6. $gcd(8, 20)$    *Answer:*

7. $gcd(9, 15)$    *Answer:*         8. $gcd(12, 16)$    *Answer:*

9. $gcd(15, 20)$    *Answer:*       10. $gcd(20, 30)$    *Answer:*

11. $gcd(12, 21)$    *Answer:*      12. $gcd(24, 32)$    *Answer:*

13. $gcd(3, 5)$    *Answer:*        14. $gcd(30, 45)$    *Answer:*

*Find the greatest common divisor of the coefficients of each binomial.*

15. $4x + 4$    *Answer:*           16. $6y + 18$    *Answer:*

17. $9a^2 + 12$    *Answer:*    18. $10t - 8$    *Answer:*

19. $20x + 25$    *Answer:*    20. $8a^2 + 20$    *Answer:*

21. $12x - 18$    *Answer:*    22. $40x + 50$    *Answer:*

**B.** *Draw a circle around the correct letter.*

23. A factor of $3x + 6$ is

    (A) 3        (B) $x$        (C) 6        (D) 9        (E) 18

24. A factor of $5y - 20$ is

    (A) 10        (B) $y$        (C) 20        (D) 5        (E) −20

25. A factor of $8a + 12$ is

    (A) 8        (B) $a$        (C) 12        (D) $a^2$        (E) 4

26. A factor of $a^2 + 7a$ is

    (A) $a^2$        (B) 7        (C) $a$        (D) $7a$        (E) 49

27. A factor of $3b^2 + 7b$ is

    (A) 3        (B) 7        (C) $b$        (D) $b^2$        (E) $3b$

28. A factor of $6x^2 + 4x$ is

    (A) $6x^2$        (B) $4x$        (C) $2x$        (D) $x^2$        (E) $4x^2$

29. A factor of $12y^3 + 15y$ is

    (A) $6y$        (B) $5y$        (C) $12y$        (D) $3y$        (E) $3y^2$

30. A factor of $4a^2b + 10ab^2$ is

    (A) $4a$        (B) $5b$        (C) $4ab$        (D) $2ab^2$        (E) $2ab$

**C.** *When several choices are given, draw a circle around the correct letter.*

31. The common factor of $4a + 16$ is

    (A) 2        (B) 4        (C) 8        (D) $8a$        (E) $4a$

32. The common factor of $10a^2 - 20a$ is

    (A) 10        (B) $a$        (C) $10a$        (D) $10a^2$        (E) $20a$

33. The common factor of $9x^2y + 6x$ is

    (A) $6x$        (B) $6xy$        (C) $3x$        (D) $3xy$        (E) $3x^2y$

34. The common factor of $15pq^2 + 25p^2q$ is

    (A) $15pq^2$        (B) $5pq^2$        (C) $5p^2q$        (D) 5        (E) $5pq$

**35.** The common factor of $16a^2b + 20a$ is _____

**36.** The common factor of $14x^2y^2 + 21xy$ is _____

**37.** Factor completely.    $6a^2 + 6a$

(A) $6a(a + 6)$          (B) $6a(a + 1)$          (C) $a(6a + 1)$

(D) $6a^2(a + 1)$        (E) $6a(a^2 + 1)$

**38.** Factor completely.    $3b^2 + 9$

(A) $3(b + 3)$          (B) $3(b^2 + 1)$          (C) $3(b^2 + 3)$

(D) $3(b^2 + 9)$        (E) $3b(b + 3)$

**39.** Factor.    $4x^2 + 8x$

(A) $4(x^2 + 2)$          (B) $4x(x^2 + 2)$          (C) $4x(x + 2)$

(D) $4x^2(x + 2)$        (E) $4x(x + 1)$

**40.** Factor.    $10a^2 - 25$

(A) $10(a^2 - 5)$          (B) $10(a^2 - 25)$          (C) $5(a^2 - 5)$

(D) $5(2a^2 - 5)$        (E) $5(2a^2 + 5)$

**41.** Factor.    $20x^3 + 30x$

**42.** Factor.    $9y^3 + 27y$          *Answer:*

**43.** Factor completely.    $2x^2y + 4xy$     *Answer:*

**44.** Factor completely.    $3a^2b^2 + 9ab$     *Answer:*

**45.** Factor completely.    $12p^2q - 18pq^2$     *Answer:*

**46.** Factor completely.    $12x^3y + 20xy^2$     *Answer:*

**47.** Factor.    $9a^2b^2 + 8a$     *Answer:*

**48.** Factor.    $8x^3y^2 + 12x^2$     *Answer:*

# 27.  Solving Equations

Equations were introduced briefly in Section 14 in order to help solve certain types of percent problems. Here we will review the basic concepts involved in order to be able to solve an equation such as

$$5x - 2 = 18$$

or

$$7x + 3 = 9x - 3$$

## A.  ROOTS OF EQUATIONS

The statement

$$3x = 6$$

is known as an *equation*. It asserts that

Three times a number is equal to 6
$$3 \qquad x \qquad = \qquad 6$$

DEFINITION

> An **equation** is a statement of equality. In the equation
>
> $$3x = 6$$
>
> $3x$ is the **left side** of the equation and 6 is the **right side**. A number is a **root** or a **solution** of an equation (in one variable) if a true statement results when that number is substituted for the variable.

Thus 2 is a root of the equation

$$3x = 6$$

because when you substitute 2 for $x$ you obtain the *true* statement

$$3(2) = 6$$

Also, 1 is *not* a root of the equation because when you substitute 1 for $x$ you obtain the *false* statement $3(1) = 6$.

The equations we will consider each have exactly one root. To check whether a number is the root of an equation, substitute the number for *each occurrence* of the variable.

**Example 1 ▶** Is 3 the root of the equation

$$4x - 2 = x + 7$$

*Solution.* Substitute 3 for each occurrence of $x$, to see whether the resulting statement is true.

$$4(3) - 2 = 3 + 7$$
$$12 - 2 = 10 \qquad true$$

Thus 3 is the root of the given equation.    ◀

**Example 2 ▶** Is –2 the root of the equation

$$2t + 1 = 5t - 5$$

*Solution.* Substitute –2 for each occurrence of $t$ to see whether the resulting statement is true.

$$2(-2) + 1 = 5(-2) - 5$$
$$-4 + 1 = -10 - 5$$
$$-3 = -15 \qquad false$$

Thus –2 is *not* the root of this equation.    ◀

*Try the exercises for Topic A on page 190.*

## B. MULTIPLICATION PROPERTY

You can think of an equation, such as

$$2x = 6$$

as a balanced scale, as in the top figure to the right. You want to isolate $x$ on one side of the equation. Because the 2 *multiplies* the $x$, and you want to isolate the unknown, you must divide the left side by 2. But now the scale is unbalanced. (See the middle figure.) If you change one side, you must do the same thing to the other side in order to preserve the balance. Thus divide both sides by 2. (See the bottom figure.) In symbols:

$$2x = 6$$

$$\frac{2x}{2} = \frac{6}{2}$$

$$x = 3$$

Thus 3 is the root of the equation $2x = 6$.

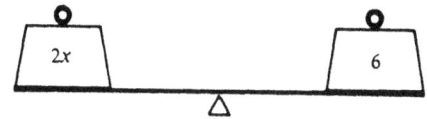

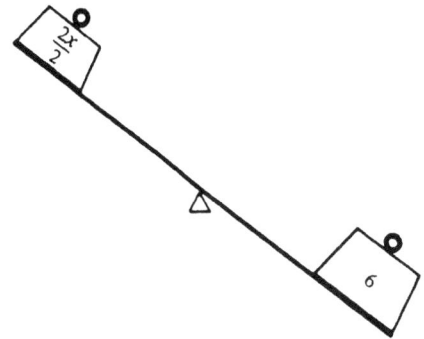

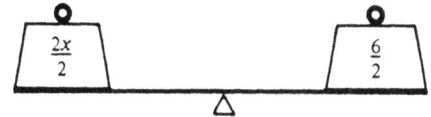

The **Multiplication Property** asserts:

> To simplify an equation, you can multiply or divide both sides by the same (nonzero) number without changing the root.

**Example 3** ▶  If $\frac{x}{3} = 4$, find $x$.

**Solution.**  What can you do to $\frac{x}{3}$ to obtain $x$?

$$\frac{x}{3} = 4 \qquad \text{Multiply both sides by 3.}$$

$$\frac{x}{3} \cdot 3 = 4 \times 3$$

$$x = 12$$

The root, $x$, is 12.

You can *check* that 12 is the root of the equation by substituting 12 for $x$ in the given equation.

$$\frac{12}{3} = 4 \qquad \textit{true}$$

The check shows that you have the correct root.  ◀

**Example 4 ▶** If $4y = 1$, find $y$.

*Solution.* What can you do to $4y$ to obtain $y$?

$$4y = 1 \qquad \text{Divide both sides by 4.}$$

$$\frac{4y}{4} = \frac{1}{4}$$

$$y = \frac{1}{4}$$

The root, $y$, is the fraction $\frac{1}{4}$.  ◀

*Try the exercises for Topic B on page 191.*

## C. ADDITION PROPERTY

The **Addition Property** asserts:

> To simplify an equation, you can add the same number to, or subtract the same number from, both sides of an equation without changing the root.

**Example 5 ▶** If $t - 3 = 4$, find $t$.

*Solution.* What can you do to $t - 3$ to obtain $t$?

$$t - 3 = 4 \qquad \text{Add 3 to both sides.}$$

$$t - 3 + 3 = 4 + 3$$

$$t = 7$$

The root, $t$, is 7.  ◀

**Example 6 ▶** If $x + 5 = 3$, find $x$.

*Solution.* What can you do to $x + 5$ to obtain $x$?

$$x + 5 = 3 \qquad \text{Subtract 5 from both sides.}$$

$$x + 5 - 5 = 3 - 5$$

$$x = -2$$

The root, $x$, is the negative integer $-2$.

*Try the exercises for Topic C on page 191.*

## D. USING BOTH PROPERTIES

You will probably have to use both the Multiplication and the Addition Properties to solve some of the equations you will be given. To simplify these equations, *bring variables to one side, numerical terms to the other.*

**Example 7 ▶**  Find $y$ if $2y + 7 = 11$.

*Solution.*

$$2y + 7 = 11 \qquad \text{Subtract 7 from both sides.}$$
$$2y + 7 - 7 = 11 - 7$$
$$2y = 4 \qquad \text{Divide both sides by 2.}$$
$$\frac{2y}{2} = \frac{4}{2}$$
$$y = 2$$

We can *check* that 2 is, indeed, the correct choice by substituting 2 for $y$ in the given equation. Thus

$$2(2) + 7 = 11$$
$$4 + 7 = 11 \qquad \textit{true} \qquad \qquad ◀$$

**Example 8 ▶**  If $5x - 1 = 2x + 8$, then $x =$

(A) 1        (B) 0        (C) $-1$        (D) 3        (E) $\frac{7}{3}$

*Solution.*  Bring variables to one side, numerical terms to the other. Because the coefficient of $x$ is larger on the left side, it is a good idea to bring the variables to the left side in this problem.

$$5x - 1 = 2x + 8 \qquad \text{Subtract } 2x \text{ from both sides.}$$
$$5x - 2x - 1 = 2x - 2x + 8$$
$$3x - 1 = 8 \qquad \text{Add 1 to both sides.}$$
$$3x - 1 + 1 = 8 + 1$$
$$3x = 9 \qquad \text{Divide both sides by 3.}$$
$$\frac{3x}{3} = \frac{9}{3}$$
$$x = 3$$

The correct choice is (D).        ◀

**Example 9 ▶**  If $3y + 2 = 4y - 7$, find $y$.

*Solution.*  Here the larger coefficient of $y$ is on the right side. Bring variables to the right side.

$$3y + 2 = 4y - 7 \qquad \text{Subtract } 3y \text{ from both sides.}$$
$$3y - 3y + 2 = 4y - 3y - 7$$
$$2 = y - 7 \qquad \text{Add 7 to both sides.}$$
$$2 + 7 = y - 7 + 7$$
$$9 = y \qquad \qquad ◀$$

**Example 10 ▶** Solve for $t$.  $\frac{t}{6} + 1 = 3$

***Solution.*** On the left side, 1 is added to $\frac{t}{6}$. Thus begin by subtracting 1 from both sides.

$$\frac{t}{6} + 1 = 3$$

$$\frac{t}{6} + 1 - 1 = 3 - 1$$

$$\frac{t}{6} = 2 \qquad \text{Multiply both sides by 6.}$$

$$\frac{t}{6} \cdot 6 = 2 \times 6$$

$$t = 12 \qquad\qquad ◀$$

*Try the exercises for Topic D on page 192.*

## EXERCISES

**A.** *Answer "yes" or "no."*

1. Is 3 the root of the equation

   $$5t = 15$$

   *Answer:*

2. Is 6 the root of the equation

   $$t + 3 = 9$$

   *Answer:*

3. Is 10 the root of the equation

   $$2x + 20 = 0$$

   *Answer:*

4. Is –3 the root of the equation

   $$\frac{y}{3} = -1$$

   *Answer:*

5. Is 4 the root of the equation

   $$5x - 4 = 12$$

   *Answer:*

6. Is 3 the root of the equation

   $$2y - 5 = 1$$

   *Answer:*

7. Is –4 the root of the equation

   $$10 - 2t = 2$$

   *Answer:*

8. Is $\frac{1}{2}$ the root of the equation

   $$6u = 3$$

   *Answer:*

9. Is 12 the root of the equation

   $$3v - 9 = 2v + 5$$

   *Answer:*

10. Is 4 the root of the equation

    $$8x - 12 = 5x + 4$$

    *Answer:*

11. Is –1 the root of the equation

    $$1 - 2y = 3 - 3y$$

    *Answer:*

12. Is 0 the root of the equation

    $$5z + 4 = 10z - 4$$

    *Answer:*

**B.** *For each equation, find the root.*

13. If $2x = 20$, find $x$.    *Answer:*        14. If $4x = 28$, find $x$.    *Answer:*

15. If $7y = 35$, find $y$.    *Answer:*        16. If $10z = 45$, find $z$.    *Answer:*

17. Find $t$ if $\frac{t}{2} = 6$.    *Answer:*        18. Find $u$ if $\frac{u}{5} = 8$.    *Answer:*

19. Find $x$ if $\frac{x}{9} = -2$.    *Answer:*        20. Find $y$ if $\frac{y}{4} = 0$.    *Answer:*

21. Solve for $t$:    $-4t = 16$        *Answer:*

22. Solve for $u$:    $-5u = -15$        *Answer:*

23. Solve for $x$:    $\frac{x}{-4} = 2$        *Answer:*

24. Solve for $y$:    $\frac{y}{-10} = -3$        *Answer:*

**C.** *For each equation, find the root.*

25. If $x + 3 = 10$, find $x$.        *Answer:*

26. If $y - 7 = 2$, find $y$.        *Answer:*

27. If $z + 4 = -2$, find $z$.        *Answer:*

28. If $u - 5 = -6$, find $u$.        *Answer:*

29. Find $x$ if $x + 8 = -3$.        *Answer:*

30. Find $y$ if $y - 7 = 0$.        *Answer:*

31. Find $t$ if $t + 6 = 0$.        *Answer:*

32. Find $u$ if $u - 8 = 8$.        *Answer:*

33. Solve for $x$:    $x + 12 = 11$        *Answer:*

34. Solve for $y$:    $9 + y = 6$        *Answer:*

35. Solve for $z$:    $-5 + z = 0$        *Answer:*

**36.** Solve for $t$:      $7 = t + 9$          *Answer:*

**D.** *When several choices are given, draw a circle around the correct letter.*

**37.** If $2x + 1 = 7$, then $x =$

(A) 8          (B) 6          (C) 4          (D) 3          (E) $\frac{7}{2}$

**38.** If $5x - 2 = 13$, then $x =$

(A) 15          (B) 11          (C) $\frac{11}{5}$          (D) $\frac{13}{5}$          (E) 3

**39.** If $6 - y = 0$, then $y =$

(A) 6          (B) –6          (C) 0          (D) 1          (E) –1

**40.** If $8 - z = 4$, then $z =$

(A) 8          (B) –8          (C) 4          (D) –4          (E) 0

**41.** Find $t$ if $4t - 3 = 17$.      *Answer:*

**42.** Find $u$ if $5u - 25 = 0$.      *Answer:*

**43.** Find $v$ if $6 - 3v = 0$.      *Answer:*

**44.** Find $x$ if $10 - 2x = 3x$.      *Answer:*

**45.** Solve for $y$:      $7y - 3 = 4y$

(A) 3          (B) $-\frac{3}{4}$          (C) 1          (D) –1          (E) –3

**46.** Solve for $z$:      $10z + 2 = 8z + 6$

(A) 2          (B) 8          (C) –2          (D) 1          (E) 0

**47.** Solve for $x$:      $6x - 3 = 9x - 9$

(A) –12          (B) –4          (C) 4          (D) 2          (E) –2

**48.** Solve for $y$:      $10 - 3y = 4 + 3y$

(A) 0          (B) 10          (C) 6          (D) 1          (E) –1

**49.** If $\frac{t}{2} - 1 = 0$, find $t$.      *Answer:*

**50.**  If $\frac{x}{3} - 1 = 5$, find $x$.          *Answer:*

**51.**  If $\frac{y}{5} + 2 = -3$, find $y$.          *Answer:*

**52.**  If $\frac{z}{6} + 8 = 2$, find $z$.          *Answer:*

**53.**  Find $x$ if $9x - 4 = 7x + 1$.          *Answer:*

**54.**  Find $y$ if $8 - 3y = 6y - 1$.          *Answer:*

**55.**  Find $t$ if $2t + 3 = 4t - 13$.          *Answer:*

**56.**  Find $u$ if $9 - 5u = 8 + u$.          *Answer:*

**57.**  Find $x$ if $6 + x = 3 - 4x$.          *Answer:*

**58.**  Find $y$ if $9y + 3 = 2 - y$.          *Answer:*

---

# 28.  Equations with Fractions

Many equations involve fractions. We begin by considering a type of equation known as a **proportion**.

## A. SOLVING PROPORTIONS

An equation, such as

$$\frac{2}{4} = \frac{1}{2}$$

that expresses the equivalence of fractions is an example of a *proportion*.

DEFINITION

A **proportion** is an equation of the form

$$\frac{a}{b} = \frac{c}{d}$$

In Example 1 the equation is a proportion.

**Example 1 ▶**  If $\frac{x}{4} = \frac{3}{5}$, find $x$.

*Solution.*  Multiply both sides by 4.

$$\frac{x}{4} = \frac{3}{5}$$

$$\frac{x}{4} \cdot 4 = \frac{3}{5} \times 4 \qquad \text{The right side equals } \frac{3}{5} \times \frac{4}{1}.$$

$$x = \frac{12}{5}$$                                                      ◀

When the variable is in the denominator, as in Example 1, a technique known as *cross-multiplying* is useful.

**Example 2 ▶**  If $\frac{4}{x} = \frac{2}{3}$, find $x$.

(A) $\frac{8}{3}$          (B) $\frac{3}{8}$          (C) $\frac{4}{3}$          (D) 12          (E) 6

*Solution.*  Cross-multiply, as indicated by the arrows.

$$\frac{4}{x} = \frac{2}{3} \qquad \text{Cross-multiply.} \quad \frac{4}{x} \times \frac{2}{3}$$

$$4 \times 3 = 2x$$

$$12 = 2x \qquad \text{Divide both sides by 2.}$$

$$\frac{12}{2} = \frac{2x}{2}$$

$$6 = x$$

The correct choice is (E).                                              ◀

*Try the exercises for Topic A on page 196.*

## B. EQUATIONS IN THE FORM OF A PROPORTION

In Examples 3 and 4, the equations, though more complicated, have the form of a proportion. Cross-multiplying is also useful in these examples.

**Example 3** ▶ If $\frac{t}{4} = \frac{t-3}{3}$, find $t$.

*Solution.*

$$\frac{t}{4} = \frac{t-3}{3}$$    Cross-multiply.    $\frac{t}{4} \bowtie \frac{t-3}{3}$

$$3t = 4(t-3)$$    Use the distributive laws on the right side.

$$3t = 4t - 12$$    Bring variables to the right side by subtracting $3t$ from both sides.

$$3t - 3t = 4t - 3t - 12$$

$$0 = t - 12$$    Add 12 to both sides.

$$0 + 12 = t - 12 + 12$$

$$12 = t$$    ◀

**Example 4** ▶ Find $x$ if $\frac{x+1}{5} = \frac{3x-4}{8}$.

*Solution.*

$$\frac{x+1}{5} = \frac{3x-4}{8}$$    Cross-multiply.    $\frac{x+1}{5} \bowtie \frac{3x-4}{8}$

$$8(x+1) = 5(3x-4)$$    Use the distributive laws.

$$8x + 8 = 15x - 20$$    Bring variables to the right side by subtracting $8x$ from both sides.

$$8x - 8x + 8 = 15x - 8x - 20$$

$$8 = 7x - 20$$    Add 20 to both sides.

$$8 + 20 = 7x - 20 + 20$$

$$28 = 7x$$    Divide both sides by 7.

$$\frac{28}{7} = \frac{7x}{7}$$

$$4 = x$$    ◀

*Try the exercises for Topic B on page 197.*

## C. MULTIPLYING BY THE *lcd*

In Examples 3 and 4, when we cross-multiplied, we actually multiplied both fractions by the *lcd* (Section 5). Some equations are not proportions, but nevertheless involve fractions. Often, multiplying both sides by the *lcd* of the fractions is the best method of attack.

**Example 5** ▶ If $\frac{x}{2} - 1 = \frac{x}{4}$, then $x =$

(A) 2          (B) 4          (C) 6          (D) 8          (E) 12

*Solution.* The *lcd* of $\frac{x}{2}$ and $\frac{x}{4}$ is 4. Multiply both sides by 4.

$$\frac{x}{2} - 1 = \frac{x}{4}$$

$$4 \cdot \left(\frac{x}{2} - 1\right) = 4 \cdot \frac{x}{4} \qquad \text{Use the distributive laws on the left side and}$$
$$\text{note that } 4 \cdot \frac{x}{2} = 2x.$$

$$2x - 4 = x$$
$$2x - x - 4 = x - x$$
$$x - 4 = 0$$
$$x - 4 + 4 = 0 + 4$$
$$x = 4$$

The correct choice is **(B)**.          ◀

**Example 6** ▶ Find $y$ if $\frac{y}{5} + 3 = \frac{y}{2}$.

*Solution.* The *lcd* of $\frac{y}{5}$ and $\frac{y}{2}$ is 10. Multiply both sides by 10.

$$\frac{y}{5} + 3 = \frac{y}{2}$$

$$10 \cdot \left(\frac{y}{5} + 3\right) = 10 \cdot \frac{y}{2} \qquad \text{Use the distributive laws and note that}$$
$$10 \cdot \frac{y}{5} = 2y.$$

$$2y + 30 = 5y \qquad \text{Subtract } 2y \text{ from both sides.}$$
$$2y - 2y + 30 = 5y - 2y$$
$$30 = 3y$$
$$\frac{30}{3} = \frac{3y}{3}$$
$$10 = y$$          ◀

*Try the exercises for Topic C on page 198.*

## EXERCISES

**A.** *In Exercises 1-6, draw a circle around the correct letter.*

1. If $\frac{x}{4} = \frac{5}{2}$, find $x$.

(A) 4            (B) 8            (C) 10            (D) $\frac{5}{8}$            (E) $\frac{4}{5}$

2.  If $\frac{x}{3} = \frac{4}{6}$, find $x$.

(A) 2            (B) 13            (C) 18            (D) $\frac{2}{9}$            (E) 12

3.  If $\frac{y}{7} = \frac{28}{4}$, find $y$.

(A) 7            (B) –7            (C) 14            (D) 35            (E) 49

4.  If $\frac{z}{10} = \frac{3}{2}$, find $z$.

(A) 5            (B) 10            (C) 15            (D) 30            (E) $\frac{3}{20}$

5.  If $\frac{t}{8} = \frac{3}{4}$, then $t =$

(A) 2            (B) 4            (C) 6            (D) 12            (E) $\frac{3}{32}$

6.  If $\frac{u}{9} = \frac{-1}{3}$, then $u =$

(A) –1            (B) –3            (C) 3            (D) –9            (E) 27

7.  If $\frac{x}{25} = \frac{2}{5}$, then $x =$      8.  If $\frac{x}{12} = \frac{-5}{6}$, then $x =$

9.  Find $y$ if $\frac{y}{3} = \frac{1}{5}$.   *Answer:*      10.  Find $x$ if $\frac{x}{4} = \frac{2}{3}$.   *Answer:*

11.  Find $t$ if $\frac{t}{9} = \frac{1}{6}$.   *Answer:*      12.  Find $x$ if $\frac{x}{12} = \frac{5}{8}$.   *Answer:*

13.  Find $x$ if $\frac{3}{x} = \frac{1}{3}$.   *Answer:*      14.  Find $y$ if $\frac{8}{y} = \frac{2}{3}$.   *Answer:*

15.  If $\frac{9}{t} = \frac{3}{2}$, find $t$.   *Answer:*      16.  If $\frac{-10}{t} = \frac{5}{2}$, find $t$.   *Answer:*

17.  If $\frac{5}{2x} = \frac{1}{4}$, find $x$.   *Answer:*      18.  If $\frac{4}{3y} = \frac{1}{8}$, find $y$.   *Answer:*

**B.** *When several choices are given, draw a circle around the correct letter.*

19.  If $\frac{x}{2} = \frac{x+3}{3}$, find $x$.

(A) 1            (B) 2            (C) 3            (D) 6            (E) 12

20.  If $\frac{y}{4} = \frac{y+4}{12}$, find $y$.

(A) 1            (B) 2            (C) 4            (D) 8            (E) 24

21. If $\frac{t}{5} = \frac{t-6}{4}$, find $t$.    *Answer:*

22. If $\frac{u}{4} = \frac{u-2}{2}$, find $u$.    *Answer:*

23. Find $x$ if $\frac{x+1}{2} = \frac{x+4}{3}$.

    (A) 2           (B) 3           (C) 4           (D) 5           (E) 15

24. Find $y$ if $\frac{y+5}{4} = \frac{y+3}{3}$.

    (A) 0           (B) 3           (C) –3          (D) 8           (E) –1

25. If $\frac{x+1}{8} = \frac{x-2}{4}$, find $x$.    *Answer:*

26. If $\frac{x-3}{7} = \frac{x+4}{14}$, find $x$.    *Answer:*

27. If $\frac{3x}{8} = \frac{x+1}{4}$, find $x$.    *Answer:*

28. Find $x$ if $\frac{5x}{4} = \frac{x+3}{2}$.    *Answer:*

29. Find $y$ if $\frac{y+3}{7} = \frac{2y-3}{5}$.    *Answer:*

30. Find $x$ if $\frac{2x+1}{3} = \frac{3x+2}{6}$.    *Answer:*

31. If $\frac{u+1}{2} = \frac{2u+2}{5}$, find $u$.    *Answer:*

32. If $\frac{v-3}{3} = \frac{v+6}{12}$, then $v =$

33. If $\frac{2x+1}{8} = \frac{4x-1}{4}$, then $x =$

34. If $\frac{x+4}{8} = \frac{3x+5}{10}$, then $x =$

**C.** *In Exercises 35-38, draw a circle around the correct letter.*

35. If $\frac{x}{2} + 1 = \frac{x}{3} + 2$, then $x =$

    (A) 6           (B) 12          (C) 24          (D) –2          (E) –12

36. If $\frac{x}{3} - 2 = \frac{x}{9}$, then $x =$

    (A) 3           (B) –3          (C) 0           (D) 9           (E) 18

37. If $\frac{x}{5} - 2 = \frac{x}{10}$, then $x =$

(A) 5     (B) –5     (C) 10     (D) 15     (E) 20

38. If $\frac{y}{9} + 1 = \frac{y}{6}$, then $y =$

(A) 3     (B) 9     (C) 18     (D) 36     (E) 54

39. Find $x$ if $\frac{x}{5} + 3 = \frac{x}{2}$.   *Answer:*

40. Find $y$ if $\frac{y}{12} + 1 = \frac{y}{10}$.   *Answer:*

41. Find $t$ if $\frac{t}{8} = \frac{t}{12} - 1$.   *Answer:*

42. Find $u$ if $\frac{3u}{4} - 1 = \frac{2u}{3}$.   *Answer:*

43. Find $x$ if $\frac{x}{2} - 6 = \frac{x}{3}$.   *Answer:*

44. Find $y$ if $\frac{y}{5} + 1 = \frac{y+1}{3}$.   *Answer:*

45. If $\frac{x}{3} - 2 = \frac{x}{5}$, then $x =$      46. If $\frac{x}{6} - 1 = \frac{x}{4}$, then $x =$

47. If $\frac{x}{3} + 1 = \frac{x}{6}$, then $x =$      48. If $\frac{x}{7} + 2 = \frac{x+4}{2}$, then $x =$

# 29.   Literal Equations

## A. SOLVING LITERAL EQUATIONS

Consider the equation

$$x + y = 10$$

There are two variables present, $x$ and $y$. At times you may want to solve this equation for $x$, and at other times, for $y$.

> A **literal equation** is one that contains at least two variables, and that must be solved for one variable in terms of the others.

For example, the literal equation

$$x + y = 10$$

can be solved for $x$ by subtracting $y$ from both sides. Thus

$$x + y - y = 10 - y$$
$$x = 10 - y$$

If you want to find $y$ in the given equation $x + y = 10$, subtract $x$ from both sides. Thus

$$x - x + y = 10 - x$$
$$y = 10 - x$$

The methods already developed for solving equations apply here as well. Use the addition and multiplication properties. If you want to solve for $x$, then isolate $x$ on one side. Thus bring all terms containing $x$ to one side, and all other terms to the other side. (You treat all other letters as though they were constants.)

If you want to solve for $y$, on the other hand, isolate $y$ on one side. Bring all terms containing $y$ to one side, and all other terms to the other side.

**Example 1 ▶**    If $4y - 5 = x$, find $y$.

*Solution.*  You want to solve for $y$. Isolate $y$ on the left side. Begin by adding 5 to both sides.

$$4y - 5 + 5 = x + 5$$

$$4y = x + 5 \qquad \text{Divide both sides by 4.}$$

$$\frac{4y}{4} = \frac{x + 5}{4}$$

$$y = \frac{x + 5}{4} \qquad\qquad\qquad\qquad ◀$$

**Example 2 ▶**    Solve for $x$.  $a = bx + 1$

*Solution.* To solve for $x$, isolate $x$ on the right side. Begin by subtracting 1 from both sides.

$$a = bx + 1$$
$$a - 1 = bx + 1 - 1$$
$$a - 1 = bx \qquad \text{Divide both sides by } b.$$
$$\frac{a - 1}{b} = x \qquad\qquad \blacktriangleleft$$

*Try the exercises for Topic A on page 203.*

## B. MULTIPLE CHOICE

Example 3 ▶ If $\frac{x + 1}{4} = y$, then $x =$

(A) $4y$  (B) $4y + 1$  (C) $4y - 1$  (D) $\frac{y}{4}$  (E) $\frac{y}{4} - 1$

*Solution.* Isolate $x$ on the left side. Begin by multiplying both sides by 4.

$$\frac{x + 1}{4} = y$$
$$4 \cdot \frac{x + 1}{4} = 4y$$
$$x + 1 = 4y$$
$$x + 1 - 1 = 4y - 1$$
$$x = 4y - 1$$

The correct choice is (C). $\qquad\qquad \blacktriangleleft$

Example 4 ▶ If $ax - b = 2c$, find $x$.

(A) $b + 2c$  (B) $\frac{b + 2c}{a}$  (C) $ab + 2c$  (D) $ab + 2ac$  (E) $\frac{2c - b}{a}$

*Solution.* Isolate $x$ on the left side. Begin by adding $b$ to both sides.

$$ax - b = 2c$$
$$ax - b + b = 2c + b$$
$$ax = 2c + b \qquad \text{Divide both sides by } a.$$
$$\frac{ax}{a} = \frac{2c + b}{a}$$
$$x = \frac{2c + b}{a}$$

Thus $x = \frac{2c + b}{a} = \frac{b + 2c}{a}$. The correct choice is (**B**). $\qquad \blacktriangleleft$

*Try the exercises for Topic B on page 203.*

## C.    APPLICATIONS

**Example 5 ▶**    The formula

$$C = \frac{5}{9}(F - 32)$$

relates the Celsius and Fahrenheit temperature scales. Here C stands for degrees Celsius and F for degrees Fahrenheit.

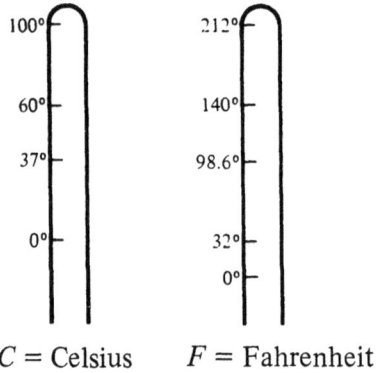

C = Celsius      F = Fahrenheit

(a)   Find C when F is 212, the boiling point of water.

(b)   Solve for F in terms of C.

(c)   Find F when C = 37.

*Solution.*

(a)   Substitute 212 for F in the given formula.

$$C = \frac{5}{9}(212 - 32)$$

$$C = \frac{5}{\cancel{9}} \cdot \overset{20}{\cancel{180}}$$
$$\phantom{C = \frac{5}{9} \cdot}{}_{1}$$

$$C = 100$$

(b)    $C = \frac{5}{9}(F - 32)$    *Divide both sides by $\frac{5}{9}$ (or multiply by $\frac{9}{5}$).*

$$\frac{9}{5}C = \frac{9}{5} \cdot \frac{5}{9}(F - 32)$$

$$\frac{9}{5}C = F - 32$$

$$\frac{9}{5}C + 32 = F$$

(c)    Substitute 37 for C in the formula obtained in part (b).

$$\frac{9}{5} \cdot 37 + 32 = F$$

$$\frac{333}{5} + 32 = F$$

$$66.6 + 32 = F$$

$$98.6 = F$$

Thus 37°C (37 degrees Celsius) corresponds to normal body temperature, 98.6°F (98.6 degrees Fahrenheit).    ◀

*Try the exercises for Topic C on page 205.*

## EXERCISES

**A.** *Solve each equation for the indicated variable.*

1.  Solve $x - y = 8$     for $x$.          *Answer:*

2.  Solve $x - y = 8$     for $y$.          *Answer:*

3.  Solve $y = 4x$     for $x$.          *Answer:*

4.  Solve $a - 3b = 0$     for $a$.          *Answer:*

5.  Solve $a - 3b = 0$     for $b$.          *Answer:*

6.  Solve $x + 2 = 3y$     for $x$.          *Answer:*

7.  Solve $2x - 1 = y$     for $x$.          *Answer:*

8.  Solve $3s + 2 = t$     for $s$.          *Answer:*

9.  Solve $5m - 3n = 1$     for $m$.          *Answer:*

10. Solve $\frac{x}{2} + 1 = y$     for $x$.          *Answer:*

11. Solve $7a - b = 2$     for $b$.          *Answer:*

12. Solve $5a - 3b = 1$     for $b$.          *Answer:*

13. Solve $\frac{1 - x}{2} = y$     for $x$.          *Answer:*

14. Solve $a + b - c = 1$     for $c$.          *Answer:*

15. Solve $ax + b = 5c$     for $x$.          *Answer:*

16. Solve $x - 2y + 3z = 0$     for $y$.          *Answer:*

17. Solve $\frac{r - s}{3} = t$     for $r$.          *Answer:*

18. Solve $10 = x + 2y + 1$     for $y$.          *Answer:*

19. Solve $\frac{1 + a}{b} = c$     for $a$.          *Answer:*

20. Solve $2x + 3y + 4z = 10$     for $x$.          *Answer:*

**B.** *Draw a circle around the correct letter.*

21. Solve $2x + y = 1$     for $x$.

    (A) $1 - y$     (B) $\frac{y - 1}{2}$     (C) $\frac{1 - y}{2}$     (D) $2 - 2y$     (E) $2y - 2$

22. Solve $5x = y + 3$     for $y$.

    (A) $\frac{5x}{3}$     (B) $5x + 3$     (C) $5x - 3$     (D) $\frac{5x - 3}{3}$     (E) $\frac{5x - 3}{5}$

23. Solve $s + 1 = 4t$    for $s$.

   (A) $4t - 1$    (B) $1 + 4t$    (C) $\frac{t}{4}$    (D) $\frac{t}{4} - 1$    (E) $4t$

24. Solve $a + 4b = 10$    for $b$.

   (A) $10 - a$    (B) $10 + a$    (C) $\frac{10 + a}{4}$    (D) $40 - a$    (E) $\frac{10 - a}{4}$

25. Solve $2x + 3y = 1$    for $y$.

   (A) $1 - 2x$    (B) $\frac{1 - 2x}{3}$    (C) $\frac{1 + 2x}{3}$    (D) $3 - 6x$    (E) $3 + 6x$

26. Solve $6u - 3v = 1$    for $v$.

   (A) $\frac{1 + 6u}{3}$    (B) $\frac{1 - 6u}{3}$    (C) $3 - 6u$    (D) $3 + 6u$    (E) $\frac{6u - 1}{3}$

27. Solve $2s + 5t = 10$    for $t$.

   (A) $10 - 2s$    (B) $\frac{10 - 2s}{2}$    (C) $\frac{10 - 2s}{5}$    (D) $\frac{10 + 2s}{5}$    (E) $50 - 10s$

28. If $4x - 9 = y$, then $x =$

   (A) $y + 9$    (B) $4y + 36$    (C) $4y - 36$    (D) $\frac{y + 9}{4}$    (E) $\frac{y}{4} + 9$

29. If $2x - y = 3$, then $x =$

   (A) $\frac{y + 3}{2}$    (B) $2y + 6$    (C) $\frac{y - 3}{2}$    (D) $\frac{y}{2} + 3$    (E) $\frac{3 - y}{2}$

30. If $a + b + 2c = 1$, then $c =$

   (A) $\frac{a - b - 1}{2}$    (B) $\frac{a + b + 1}{2}$    (C) $\frac{1 - a - b}{2}$    (D) $2 - 2a - 2b$    (E) $2 + 2a + 2b$

31. If $ax + 2b = c$, then $x =$

   (A) $\frac{2b - c}{a}$    (B) $\frac{c - 2b}{a}$    (C) $\frac{2b + c}{a}$    (D) $2ab + ac$    (E) $ac - 2b$

32. If $r - 2s + 4t = 0$, then $t =$

   (A) $2s - r$    (B) $8s - r$    (C) $8s - 4r$    (D) $\frac{2s - r}{4}$    (E) $\frac{r - 2s}{4}$

33. If $3x - y + 5z = 1$, then $x =$

   (A) $1 + y - 5z$    (B) $\frac{1 + y + 5z}{3}$    (C) $\frac{1 + y - 5z}{3}$    (D) $3 + y - 5z$    (E) $\frac{5z - y + 1}{3}$

34. If $a - 4c = 2b$, then $c =$

   (A) $\frac{2b - a}{4}$    (B) $\frac{a - 2b}{4}$    (C) $4a - 8b$    (D) $8b - a$    (E) $\frac{a - 4b}{2}$

35.  If $\frac{3x - 1}{4} = y$, then $x =$

(A) $4y + 1$    (B) $12y + 4$    (C) $\frac{4y - 1}{3}$    (D) $\frac{4y + 1}{3}$    (E) $\frac{4y}{3} + 1$

36.  If $\frac{1 + 2y}{5} = 3x$, then $y =$

(A) $\frac{3x + 5}{2}$    (B) $\frac{3x - 5}{2}$    (C) $15x - 1$    (D) $\frac{15x - 1}{2}$    (E) $30x - 2$

C.

37.  (a)  The formula

$$A = lw$$

expresses the area, $A$, of a rectangle in terms of the length, $l$, and width, $w$. Solve for $l$.

(b)  Find $l$ when $A = 28$ square meters and $w = 4$ meters.

*Answer:*    (a)                (b)

38.  (a)  The formula

$$P = 2l + 2w$$

expresses the perimeter, $P$, of a rectangle in terms of the length, $l$, and the width, $w$. Solve for $l$.

(b)  Find $l$ when $P = 80$ feet and $w = 12$ feet.

*Answer:*    (a)                (b)

39.  (a)  The formula

$$A = \frac{bh}{2}$$

expresses the area, $A$, of a triangle in terms of $b$, the length of the base, and $h$, the length of the altitude. Solve for $h$.

(b)  Find $h$ when $A = 120$ square inches and $b = 8$ inches.

*Answer:*    (a)                (b)

40.  (a)  The formula

$$V = lwh$$

expresses the volume, $V$, of a rectangular box in terms of the length, $l$, the width, $w$, and the height, $h$. Solve for $h$.

(b)  Find $h$ when $V = 4000$ cubic centimeters, $l = 20$ centimeters, and $w = 10$ centimeters.

Answer:     (a)                    (b)

41. (a)  The formula

$$d = rt$$

expresses the distance, $d$, that can be traveled at a constant rate, $r$, over a period of time, $t$. Solve for $r$.

(b)  Find $r$ when $d = 600$ kilometers and $t = 5$ hours.

Answer:     (a)                    (b)

42. (a)  The formula

$$R = \frac{kl}{t^2}$$

expresses the resistance, $R$, of an electrical wire in terms of the length, $l$, of the wire, the thickness, $t$, of the wire, and a numerical factor, $k$. Solve for $l$.

(b)  Suppose the resistance is 80 ohms when the wire is .2 inch thick and when $k = \frac{1}{40,000}$. Find the length of the wire.

Answer:     (a)                    (b)

43. The formula

$$R \cdot P \cdot T = I$$

expresses the interest, $I$, earned when a sum of money is invested over a period of time. Here $R$ is the interest rate, $P$, is the sum of money, called the *principal*, and $T$ is a period of time, expressed as a fraction of a year.

(a)  Solve for $P$.

(b)  Find $P$ when $R = 8\%$, $T = \frac{3}{4}$, and $I = \$1200$.

Answer:             (a)                    (b)

# 30.   *x, y*-Plane

In order to discuss the graph of an equation such as

$$y = 2x + 3$$

you must first know how *pairs* of numbers correspond to points on a plane.

## A.  *x*-AXIS

There is a convenient way of picturing numbers along a horizontal line, which will be called the *x*-**axis**. Suppose that the *x*-axis extends without limit both to the left and to the right. Choose any point on this line and call it the **origin**. This point corresponds to the number 0. Now choose another point to the *right* of 0. This second point corresponds to the number 1.

The length of the segment of the line between 0 and 1 determines one **unit of distance**. This is represented by the thickened portion of the line below. The number 2 corresponds to the point one (distance) unit to the right of 1. The number 3 corresponds to the point one unit further to the right. Continue this process to obtain the numbers 4, 5, 6, and so on, as in the figure below.

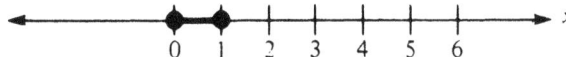

Positive numbers are pictured to the *right* of 0 along the number line. It seems reasonable that negative numbers should be pictured to the *left* of 0. Thus go one distance unit to the *left* of 0. This point corresponds to −1. Go one unit further to the left to locate −2. Continue to the left to locate −3, −4, −5, and so on, as in the figure below.

Observe that $3 < 5$ and 3 lies to the left of 5 on the *x*-axis. In general,

$$a < b \text{ when } a \text{ lies to the } left \text{ of } b$$

Thus $-5 < -3$ because −5 lies to the left of −3.

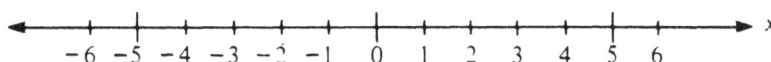

Midway between 0 and 1 is the point corresponding to the fraction $\frac{1}{2}$. (See the figure below.) The point corresponding to $\frac{7}{4}$, that is, to $1\frac{3}{4}$, lies three-fourths of the way from 1 to 2. Negative fractions lie to the left of 0. Thus $-\frac{1}{2}$ lies midway between $-1$ and 0.

**Example 1 ▶** Consider the $x$-axis, pictured below. Label the points that correspond to the following numbers.

(a)  4            (b) 7            (c)  $-1$            (d)  $-5$            (e)  $\frac{3}{2}$

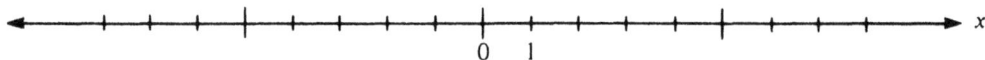

***Solution.*** In part (e), note that $\frac{3}{2} = 1\frac{1}{2}$. Thus $\frac{3}{2}$ lies midway between **1 and 2.** See the figure below.

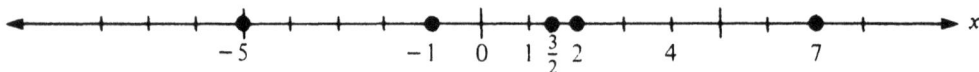

*Try the exercises for Topic A on page 210.*

### B.  $y$-AXIS

Draw a vertical line through the origin, 0, on the $x$-axis. This vertical line will be called the **$y$-axis**. The two lines intersect at the origin, which will correspond to 0 on the $y$-axis, as well as on the $x$-axis. Use the same distance unit on the $y$-axis as on the $x$-axis.
Positive numbers are located *upward* from 0 on the $y$-axis, and negative numbers *downward*, as indicated in the figure below.

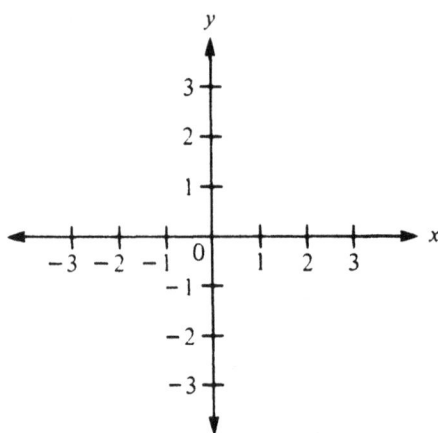

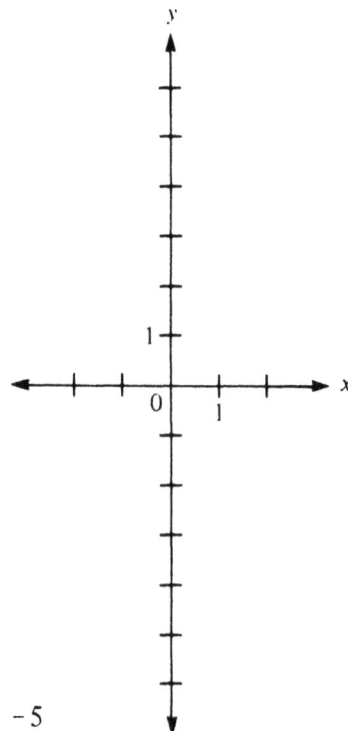

**Example 2 ▶** In the figure to the right, label the points on the $y$-axis that correspond to the following numbers.

(a)  $\frac{1}{2}$            (b)  6            (c)  $-2$            (d)  $-5$

**Solution.**

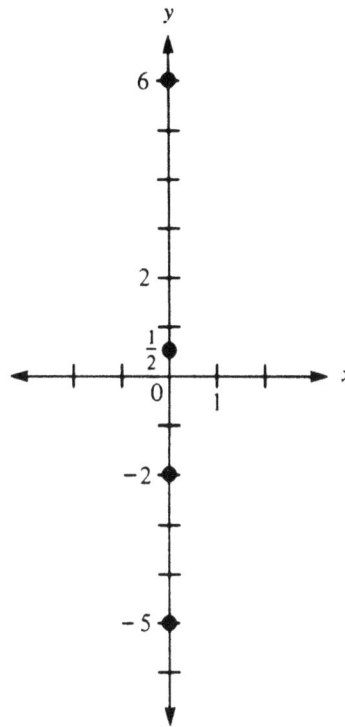

*Try the exercises for Topic B on page 211.*

### C. COORDINATES

Every number corresponds to a point on the *x*-axis and to a point on the *y*-axis.  If we consider two numbers, *a* and *b*, then *a* corresponds to a point on the *x*-axis and *b* to a point on the *y*-axis. Locate these points and draw lines perpendicular to the *x*- and *y*-axes through each of them, as in the figure below to the left. The intersection of these perpendicular lines is the point *P* on the "*x,y*-plane corresponding to the numbers *a* and *b* *in this order*. We write

$$P = (a, b)$$

The number *a* is called the **x-coordinate** of *P*, and the number *b* is the **y-coordinate**. Together, *a* and *b* are called the **coordinates** of *P*.

For example, 2 is the *x*-coordinate of (2, 4) and 4 is the *y*-coordinate. On the other hand, if we change the order of the coordinates, then 4 is the *x*-coordinate and 2 is the *y*-coordinate of (4, 2). Note that

$$(2, 4) \neq (4, 2)$$

They are *different* points. See the figure at the right.

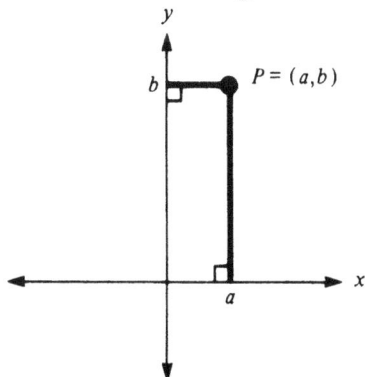

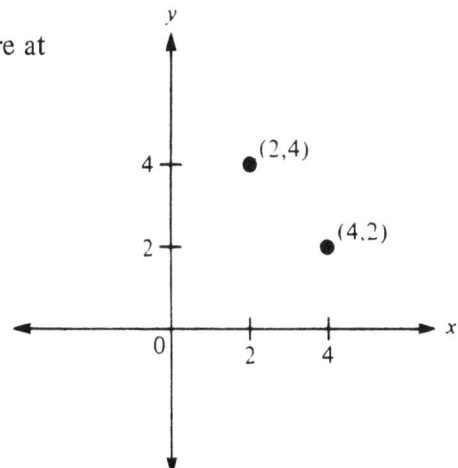

**Example 3** ▶   For each point in the figure below, find:

    (a) its $x$-coordinate,

    (b) its $y$-coordinate.

    (c) Express the point in terms of its coordinates.

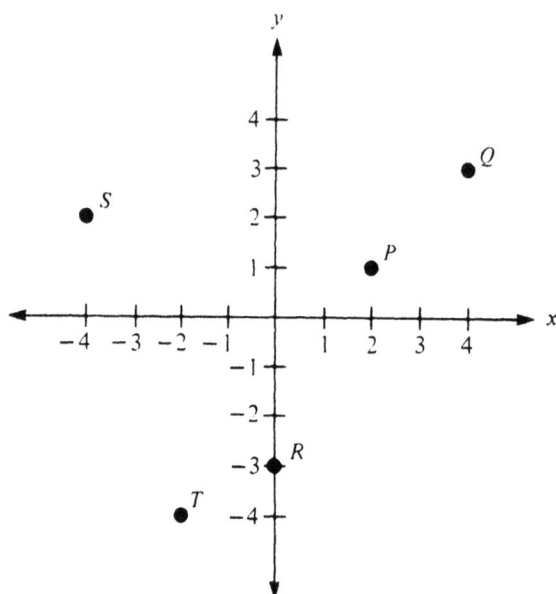

*Solution.*

$P$:  (a) 2    (b) 1    (c) $P = (2, 1)$

$Q$:  (a) 4    (b) 3    (c) $Q = (4, 3)$

$R$:  (a) 0    (b) -3   (c) $R = (0, -3)$

$S$:  (a) -4   (b) 2    (c) $S = (-4, 2)$

$T$:  (a) -2   (b) -4   (c) $T = (-2, -4)$              ◀

*Try the exercises for Topic C on page 211.*

## EXERCISES

**A.** *Consider the x-axis, pictured below. Label the points that correspond to the following numbers.*

1. 3           2. 6           3. -2           4. -4

5. -6        6. $\dfrac{1}{2}$        7. $\dfrac{5}{2}$        8. $-1\dfrac{1}{2}$

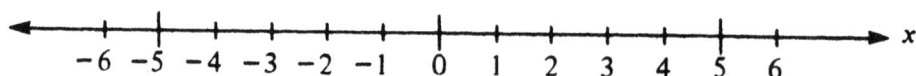

**B.** *Consider the y-axis, pictured at the right. Label the points that correspond to the following numbers.*

9.  4                  10.  7

11.  −1                12.  −3

13.  −6               14.  $\frac{3}{2}$

15.  $3\frac{1}{2}$            16.  $-\frac{1}{2}$

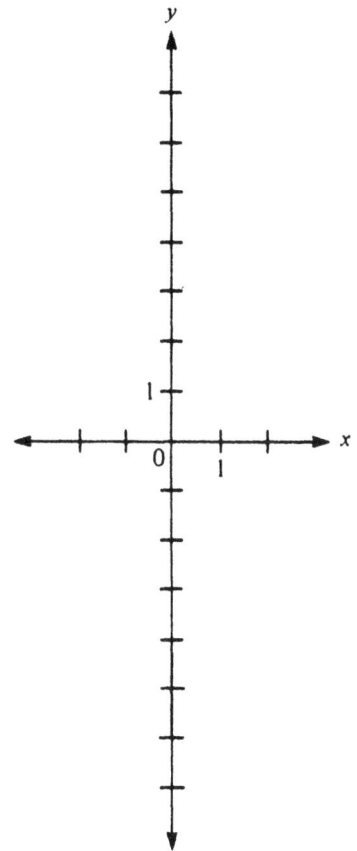

**C.** *For each point in the figure at the right, find:*

  (a)  its *x*-coordinate.

  (b)  its *y*-coordinate.

  (c)  Express the point in terms of its coordinates.

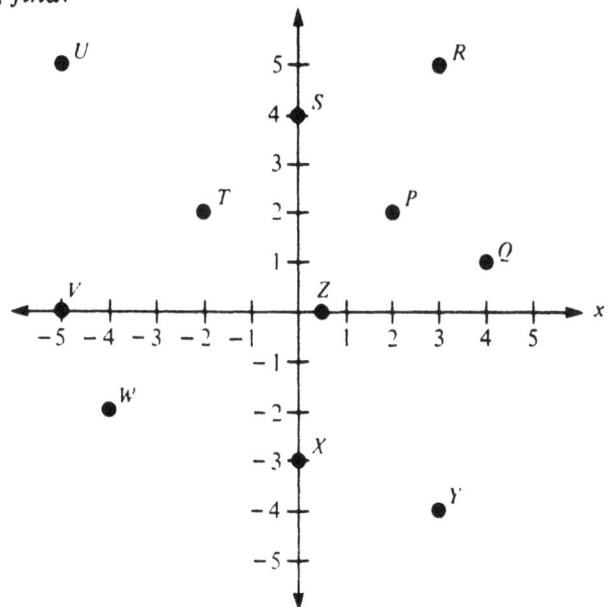

17.  P        *Answer:*    (a)              (b)              (c)  P =

18.  Q        *Answer:*    (a)              (b)              (c)  Q =

19.  R        *Answer:*    (a)              (b)              (c)  R =

20.  S        *Answer:*    (a)              (b)              (c)  S =

21.  T        *Answer:*    (a)              (b)              (c)  T =

22.  U            *Answer:*      (a)              (b)              (c) U =

23.  V            *Answer:*      (a)              (b)              (c) V =

24.  W            *Answer:*      (a)              (b)              (c) W =

25.  X            *Answer:*      (a)              (b)              (c) X =

26.  Y            *Answer:*      (a)              (b)              (c) Y =

27.  Z            *Answer:*      (a)              (b)              (c) Z =

*In the figure below, locate the following points.*

28.  P = (1, 3)

29.  Q = (0, 5)

30.  R = (−2, 1)

31.  S = (−5, 5)

32.  T = (−3, 0)

33.  U = (−4, −3)

34.  V = (0, −2)

35.  W = (−4, 2)

36.  X = (2, −4)

37.  Y = (3, 0)

38.  $Z = \left(\dfrac{9}{2}, 0\right)$

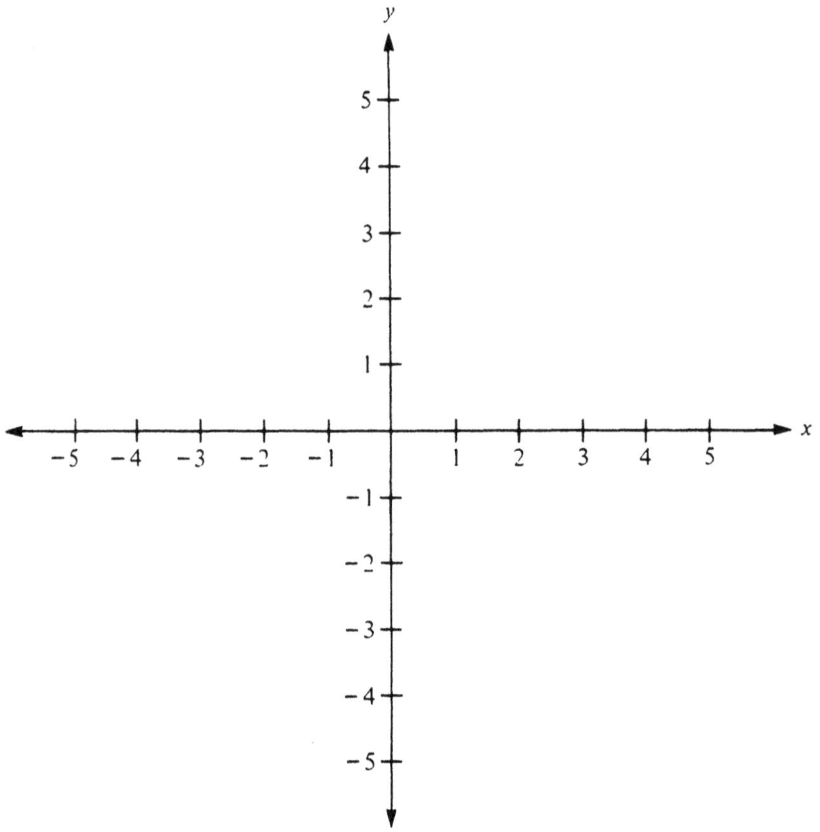

# 31.  Graph of a Line

How can we represent an equation on the $x,y$-plane?

## A.  DRAWING A LINE

We will examine equations in *two* variables, $x$ and $y$, such as

$$y = x + 3$$

and

$$y = 2x - 1$$

First, consider the equation

$$y = x + 3$$

Observe that when we substitute 1 for $x$ and 4 for $y$, we obtain a *true* statement,

$$4 = 1 + 3$$

Now consider the point $(1, 4)$ with $x$-coordinate 1 and $y$-coordinate 4. We will say that $(1, 4)$ is a *solution* of the equation $y = x + 3$.

DEFINITION

> The point $(a, b)$ is a **solution** of an equation in the variables $x$ and $y$ if a true statement results when $a$ is substituted for $x$ and $b$ for $y$.

**Example 1 ▶**  (a)  Is $(2, 5)$ a solution of the equation $y = x + 3$?

(b)  Is $(5, 10)$ a solution of the equation $y = x + 3$?

*Solution.*  Consider the equation

$$y = x + 3$$

(a)  Substitute 2 for $x$ and 5 for $y$.

$$5 = 2 + 3 \qquad (\textit{true})$$

The point $(2, 5)$ is a solution of the equation $y = x + 3$.

(b) Substitute 5 for $x$ and 10 for $y$.

$$10 = 5 + 3 \qquad (false)$$

The point $(5, 10)$ is *not* a solution of the equation $y = x + 3$.    ◀

**DEFINITION**

> The **graph** of an equation in $x$ and $y$ is its pictorial representation on the $x, y$-plane. The graph consists of all points $(x, y)$ that are solutions of the equation.

We already know that the points $(1, 4)$ and $(2, 5)$ are solutions of the equation $y = x + 3$. Thus each of these points is *on* the graph of $y = x + 3$.

**Example 2** ▶    Graph the equation $y = x + 3$.

*Solution.* Choose several values of $x$ and find the corresponding values of $y$, as in the following table.

| $x$ | $x + 3$ | $y$ |
|---|---|---|
| 0 | $0 + 3$ | 3 |
| 1 | $1 + 3$ | 4 |
| 2 | $2 + 3$ | 5 |
| 3 | $3 + 3$ | 6 |
| 4 | $4 + 3$ | 7 |

Locate the corresponding points $(x, y)$ on the $x, y$-plane, as in the figure at the right. Observe that a straight line appears to go through these points.

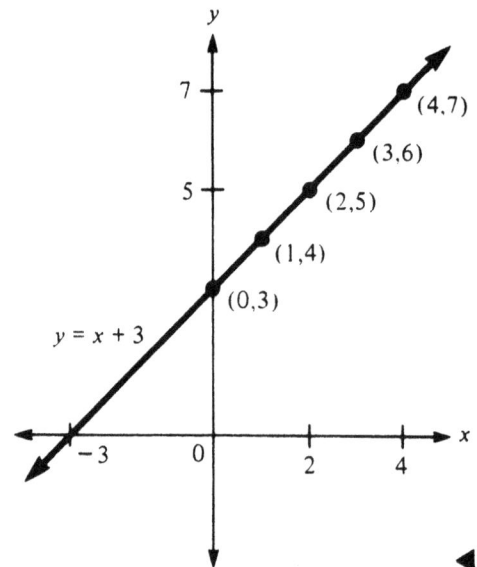

The graph of an equation such as

$$y = x + 3, \qquad y = 2x, \qquad y = 1 - x, \qquad y = 2, \qquad \text{or} \qquad x = 3$$

is a line. In such an equation, $x$ or $y$ or both of these variables may appear, and *only to the first power*. Thus the graph of

$$y = x^2$$

is *not* a straight line (because the *second* power of $x$ appears).

*A line is determined by two points.* A line extends without limit in two directions.

> To draw a line:
>
> 1. Choose two convenient values of $x$. (It is often easiest to choose $x = 0$ and $x = 1$.)
> 2. Find the corresponding values of $y$.
> 3. Plot the corresponding points $(x, y)$.
> 4. Draw the line connecting these two points.

(In the case of a vertical line, Step 1 will be modified in *Topic* C.)

**Example 3** ▶   Graph the equation $y = 2x$.

**Solution.** In this equation $x$ and $y$ appear only to the first power. Consequently, this is the equation of a line.

1. Choose two values of $x$. We will take $x = 0$ and $x = 1$.
2. The corresponding values of $y$ are given in the table below.

| $x$ | $2x$ | $y$ |
|---|---|---|
| 0 | $2 \times 0$ | 0 |
| 1 | $2 \times 1$ | 2 |

3. Locate the two points $(0, 0)$ and $(1, 2)$ on the $x, y$-plane.

4. Draw the line connecting these points, as in the figure at the right. The arrowheads indicate that the line extends without limit in both directions.

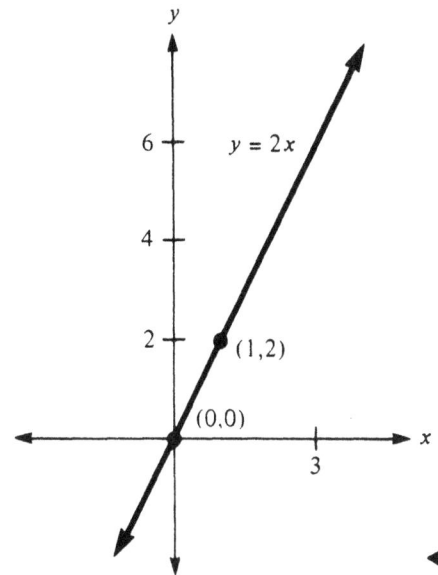

**Example 4** ▶   Graph the equation $y = 1 - x$.

**Solution.** This is the equation of a line because $x$ and $y$ appear only to the first power.

1. Let $x = 0$ and let $x = 1$.
2. Find the corresponding values of $y$, as in the table below.

| $x$ | $1 - x$ | $y$ |
|---|---|---|
| 0 | $1 - 0$ | 1 |
| 1 | $1 - 1$ | 0 |

3. Locate the points $(0, 1)$ and $(1, 0)$ on the $x, y$-plane.

4. Draw the line connecting these points, as in the figure at the right.

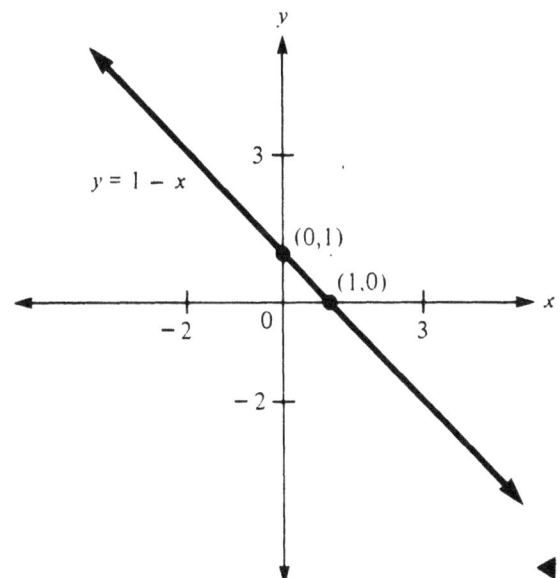

*Try the exercises for Topic A on page 218.*

### B. FINDING POINTS ON A LINE

**Example 5 ▶**  Which of these points lies on the graph of $3x + y = 5$?

(A) $(1, 1)$     (B) $(2, 3)$     (C) $(2, 1)$     (D) $(1, 2)$     (E) $(3, 2)$

*Solution.* Check each point by substituting its $x$-coordinate for $x$ and its $y$-coordinate for $y$ in the equation

$$3x + y = 5$$

Begin with choice (A) and continue until you obtain a true statement.

(A)    Substitute 1 for $x$ and 1 for $y$ in the equation $3x + y = 5$.

$$3(1) + 1 = 5$$
$$3 + 1 = 5 \qquad \textit{false}$$

(B)    Substitute 2 for $x$ and 3 for $y$.

$$3(2) + 3 = 5$$
$$6 + 3 = 5 \qquad \textit{false}$$

(C)    Substitute 2 for $x$ and 1 for $y$.

$$3(2) + 1 = 5$$
$$6 + 1 = 5 \qquad \textit{false}$$

(D)    Substitute 1 for $x$ and 2 for $y$.

$$3(1) + 2 = 5$$
$$3 + 2 = 5 \qquad \textit{true}$$

The correct choice is (D).                                        ◀

*Try the exercises for Topic B on page 219.*

### C. HORIZONTAL LINES AND VERTICAL LINES

Every point on a *horizontal line* has the same $y$-coordinate. The figure to the right is the graph of the horizontal line

$$y = 2$$

Among the points on this line are

$(0, 2)$,    $(1, 2)$,
$(2, 2)$,    $(-1, 2)$

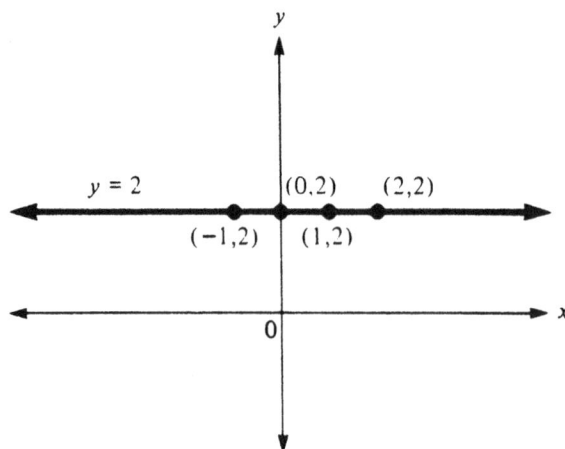

Other horizontal lines are given by equations such as

$$y = 5$$

$$y = 0 \qquad \textit{This is the equation of the x-axis.}$$

$$y = -1$$

Every point on a *vertical line*, has the same $x$-coordinate. (Thus when graphing a vertical line, choose two different $y$-values corresponding to this $x$-value.) The figure to the right is the graph of the vertical line

$$x = 3$$

Among the points on this line are

$$(3, 0), \quad (3, 1), \quad (3, 3), \quad (3, -3)$$

Other vertical lines are given by equations such as

$$x = 4$$

$$x = 0 \qquad \textit{This is the equation of the y-axis.}$$

$$x = -3$$

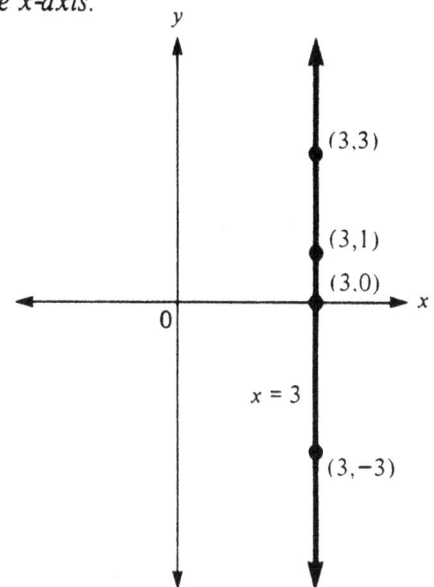

**Example 6** ▶   (a)  Which equations represent horizontal lines?
(b)  Which equations represent vertical lines?

(A) $y = 4$ \qquad (B) $x = -2$ \qquad (C) $x + y = 2$

(D) $y - 1 = 0$ \qquad (E) $y = x - 3$

*Solution.*

(a)  Equations (A) and (D) represent horizontal lines. Each can be written in the form

$$y = \text{some number}$$

Observe that in (D), you can add 1 to each side to obtain

$$y - 1 + 1 = 0 + 1$$

or

$$y = 1$$

(b)  Equation (B) is of the form

$$x = \text{some number}$$

and thus represents a vertical line.  ◀

*Try the exercises for Topic C on page 220.*

**EXERCISES**

**A.** *Indicate which points are solutions of the given equations.*

1.  Is $(1, 6)$ a solution of the equation $y = x + 5$?          *Answer:*

2.  Is $(2, 4)$ a solution of the equation $y = 2x$?          *Answer:*

3.  Is $(4, 3)$ a solution of the equation $y = x - 1$?          *Answer:*

4.  Is $(1, 1)$ a solution of the equation $y = -x$?          *Answer:*

5.  Is $(8, 2)$ a solution of the equation $y = 4x$?          *Answer:*

6.  Is $(1, 5)$ a solution of the equation $y = 2x + 1$?          *Answer:*

7.  Is $(0, 0)$ a solution of the equation $2y = 3x$?          *Answer:*

8.  Is $(3, 1)$ a solution of the equation $x + y = 4$?          *Answer:*

9.  Is $(1, 2)$ a solution of the equation $2x - y = 1$?          *Answer:*

10.  Is $(5, 3)$ a solution of the equation $y = 3$?          *Answer:*

11.  Is $(2, -4)$ a solution of the equation $y = 2$?          *Answer:*

12.  Is $(1, -1)$ a solution of the equation $x = 1$?          *Answer:*

*In Exercises 13–22 graph each equation.*

13.  Graph the equation   $y = x + 1$.          14.  Graph the equation   $y = x - 1$.

15.  Graph the equation   $y = x + 4$.          16.  Graph the equation   $y = x - 2$.

17. Graph the equation   $y = 3x$.

18. Graph the equation   $y = -2x$.

19. Graph the equation   $y = 2 - x$.

20. Graph the equation   $y = 2x + 1$.

21. Graph the equation   $y = 1 - 2x$.

22. Graph the equation   $y = 4 - 2x$.

B. *Draw a circle around the correct letter.*

23. Which of these points lies on the graph of $x + y = 6$?

(A) $(3, 2)$          (B) $(2, 2)$          (C) $(2, 4)$          (D) $(6, 1)$          (E) $(0, -6)$

24. Which of these points lies on the graph of $y = x + 8$?

(A) $(0, 8)$          (B) $(4, 4)$          (C) $(8, 0)$          (D) $(0, -8)$          (E) $(2, 6)$

25. Which of these points lies on the graph of $y = 2x + 5$?

(A) $(0, 0)$          (B) $(1, 5)$          (C) $(2, 5)$          (D) $(2, 1)$          (E) $(0, 5)$

26. Which of these points lies on the graph of $y = 3x - 1$?

   (A) $(1, 1)$     (B) $(2, 5)$     (C) $(5, 2)$     (D) $(-1, -1)$     (E) $(3, 1)$

27. Which of these points lies on the graph of $y = 3 - x$?

   (A) $(0, 3)$     (B) $(3, 3)$     (C) $(3, 1)$     (D) $(-3, 0)$     (E) $(2, 2)$

28. Which of these points lies on the graph of $x + y = -1$?

   (A) $(6, 5)$     (B) $(6, -5)$     (C) $(-6, 5)$     (D) $(0, 0)$     (E) $(-1, 4)$

29. Which of these points lies on the graph of $2x + 3y = 10$?

   (A) $(2, 3)$     (B) $(2, 2)$     (C) $(3, 3)$     (D) $(3, 2)$     (E) $(10, 0)$

30. Which of these points lies on the graph of $4x + y = 0$?

   (A) $(1, 4)$     (B) $(4, 1)$     (C) $(1, -4)$     (D) $(-1, -1)$     (E) $(-1, -4)$

31. Which of these points lies on the graph of $4x + 3y = 12$?

   (A) $(3, 4)$     (B) $(4, 3)$     (C) $(2, 1)$     (D) $(7, -2)$     (E) $(6, -4)$

32. Which of these points lies on the graph of $2x + 5y = 3$?

   (A) $(2, -1)$     (B) $(4, -1)$     (C) $(-1, 4)$     (D) $\left(\frac{1}{2}, 1\right)$     (E) $(7, -2)$

33. Which of these points lies on the graph of $4x + 3y = 7$?

   (A) $\left(\frac{1}{2}, 2\right)$     (B) $\left(\frac{1}{4}, -2\right)$     (C) $\left(2, \frac{1}{3}\right)$     (D) $\left(2, -\frac{1}{3}\right)$     (E) $(4, -2)$

34. Which of these points lies on the graph of $2x - 5y = 4$?

   (A) $(6, 2)$     (B) $(3, 1)$     (C) $(5, 1)$     (D) $(7, 2)$     (E) $(10, 3)$

C. (a) Which of these equations represent horizontal lines?

   (b) Which represent vertical lines?

   *Answer:* (a), (b), *or* neither

35. $y = 8x$          *Answer:*          36. $y = 8$          *Answer:*

37. $x = 8$          *Answer:*          38. $x = y + 8$          *Answer:*

39. $x + 8 = 0$          *Answer:*          40. $8x - 8 = 0$          *Answer:*

41. $x + y = 8$          *Answer:*          42. $y + 8 = 0$          *Answer:*

**43.** Graph the equation   $y = 5$.

**44.** Graph the equation   $x = 5$.

*Draw a circle around the correct letter.*

**45.** Which of these points lies on the line $y = -3$?

   (A) $(-3, 0)$      (B) $(-3, 3)$      (C) $(-1, -2)$      (D) $(0, -3)$      (E) $(0, 0)$

**46.** Which of these points lies on the line $x = 10$?

   (A) $(-10, 10)$    (B) $(0, 10)$      (C) $(0, -10)$      (D) $(5, 5)$      (E) $(10, 5)$

# 32.  Equation of a Line

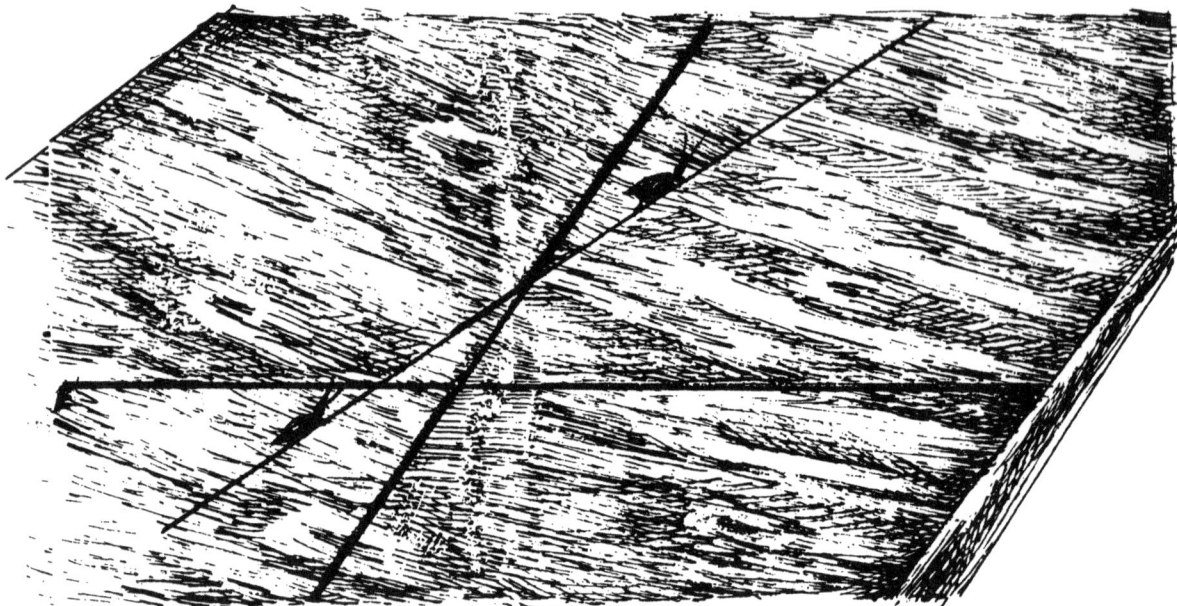

Often, to answer a question about a line, you must rewrite its equation in a special form. Here we consider two such special forms.

## A.  SLOPE-INTERCEPT FORM

The equation of a line that is not vertical can be written in the form

$$y = mx + b$$

For example,

$$y = 2x + 3$$

is of this form. Here $m = 2$ and $b = 3$.

The equation

$$x + 2y = 1$$

can be written in the form $y = mx + b$ by treating $x + 2y = 1$ as a literal equation and solving for $y$:

$$x - x + 2y = 1 - x$$

$$2y = 1 - x$$

$$\frac{2y}{2} = \frac{1 - x}{2} \qquad \text{On the right side, write the } x\text{-term first.}$$

$$y = -\frac{1}{2}x + \frac{1}{2}$$

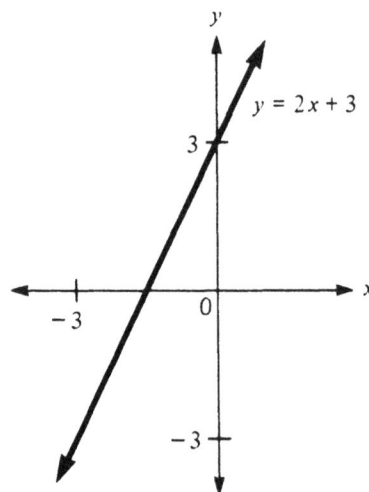

Thus $m = -\frac{1}{2}x$ and $b = \frac{1}{2}$

In the graph of $y = 2x + 3$, observe that the line crosses the $y$-axis where $y = 3$.

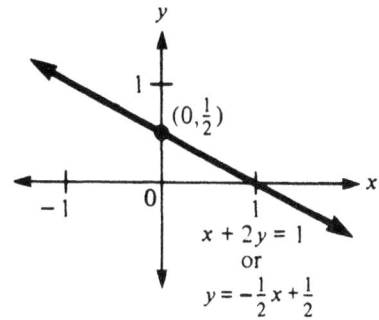

DEFINITION

> The $y$-intercept of a line is the $y$-coordinate of its intersection with the $y$-axis.

Thus 3 is the $y$-intercept of the line with equation $y = 2x + 3$.

To find the $y$-intercept of a line, note that $x$ must be 0 where the line intersects the $y$-axis. Thus substitute 0 for $x$ and solve for $y$. When the equation of the line is of the form

$$y = mx + b$$

substitute 0 for $x$ and obtain

$$y = 0 + b$$

so that

$$y = b$$

Thus $b$ *represents the y-intercept.*

Example 1 ▶ At what point does the graph of $y = x + 5$ intersect the $y$-axis?

*Solution.* This equation is of the form $y = mx + b$. Here $b = 5$. Thus 5 is the $y$-intercept. The graph intersects the $y$-axis at $(0, 5)$. See the figure to the right.

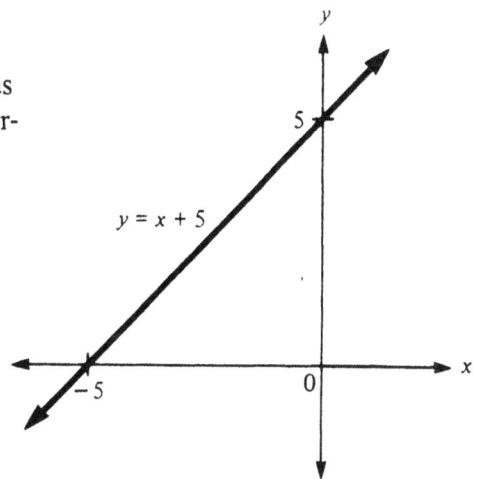

Example 2 ▶ If the point $(0, b)$ lies on the graph of

$$y = 3x - 1 \text{ then } b =$$

***Solution.*** If the point $(0, b)$ lies on the line, then $b$ is the $y$-intercept of the line. The given equation is of the form $y = mx + b$, where $b$, the $y$-intercept, equals $-1$.

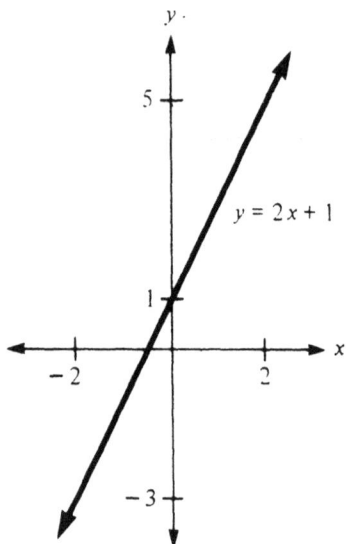

$y = 3x - 1$

$y = 2x + 1$

In the equation

$$y = mx + b$$

the number $m$ is called the **slope** of the line. *When $m$ is positive, the line slopes upward to the right,* as in the figure to the left, which represents the line with equation

$$y = 2x + 1$$

*When $m$ is negative, the line slopes downward to the right* (or equivalently, *upward to the left,* as in the figure below to the right, which represents the line with equation

$$y = -2x + 1$$

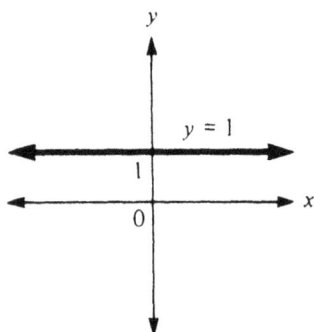

*When $m = 0$, the line is horizontal,* as in the figure to the left, which represents the line with equation

$$y = 1$$

$y = 1$

$y = -2x + 1$

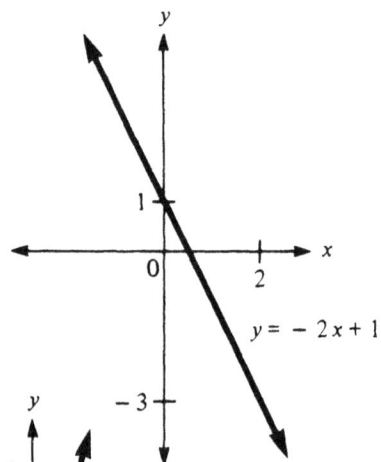

When the absolute value of $m$ is "large," the line rises steeply, as in the figure to the right, which represents the line with equation

$$y = 4x + 1 \qquad (m = 4)$$

(Think of a *steep* hill.)

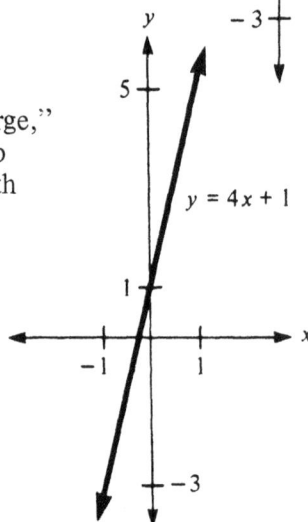

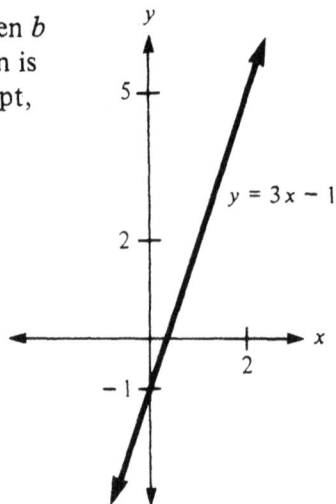

$y = 4x + 1$

When the absolute value of $m$ is "small," the line rises gradually, as in the figure below, which represents the line with equation

$$y = \frac{1}{3}x \qquad \left(m = \frac{1}{3}\right)$$

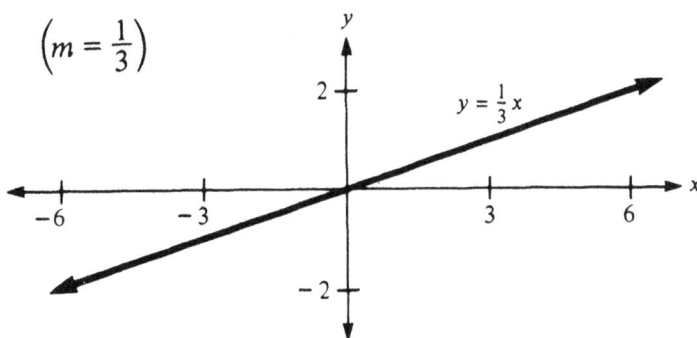

The equation

$$y = mx + b$$

is known as the **slope-intercept form** of the equation of a line.

*Try the exercises for Topic A on page 228.*

## B. FINDING AN EQUATION FROM A GRAPH

A line intersects the $x$-axis when $y = 0$. Thus to find this intersection, substitute 0 for $y$ in the given equation of the line. For the line

$$y = 2x + 4$$

substitute 0 for $y$, and solve for $x$.

$$0 = 2x + 4$$
$$0 - 4 = 2x + 4 - 4$$
$$-4 = 2x$$
$$\frac{-4}{2} = \frac{2x}{2}$$
$$-2 = x$$

Thus the line intersects the $x$-axis at $(-2, 0)$.

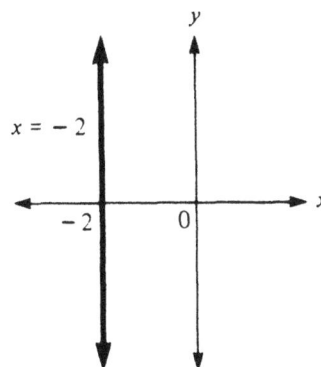

DEFINITION

> The *x*-**intercept** of a line is the $x$-coordinate of its intersection with the $x$-axis.

Suppose you know that $a$ is the $x$-intercept and that $b$ is the $y$-intercept of a line. Then the equation of the line can be written in the form

$$\frac{x}{a} + \frac{y}{b} = 1$$

This is known as the **intercept form** of the equation of a line. For example, if the x-intercept of a line is 2 and the y-intercept is 3, then the line has the equation

$$\frac{x}{2} + \frac{y}{3} = 1$$

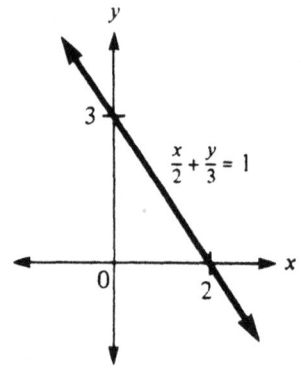

When you are given the graph of a line, it is often easiest to find its equation in intercept form. See where the line crosses the x-axis to find the x-intercept, a, and see where the line crosses the y-axis to find the y-intercept, b. Write the equation in intercept form

$$\frac{x}{a} + \frac{y}{b} = 1$$

Sometimes, you may want to multiply both sides by the lcd, ab. Thus, for the equation

$$\frac{x}{2} + \frac{y}{3} = 1$$

which is in intercept form, multiply both sides by 6, the lcd of $\frac{x}{2}$ and $\frac{y}{3}$.

$$6\left(\frac{x}{2} + \frac{y}{3}\right) = 6 \times 1$$

Use the distributive laws on the left side to obtain

$$6 \cdot \frac{x}{2} + 6 \cdot \frac{y}{3} = 6$$

and finally,

$$3x + 2y = 6$$

**Example 3 ▶** An equation for the line at the right is

(A) $x = 3$        (B) $y = 3$

(C) $x + y = 3$     (D) $x - y = 3$

(E) $y = x + 3$

***Solution.*** Clearly, the x-intercept is 3 and the y-intercept is also 3. Thus $a = 3$ and $b = 3$. The intercept form of the equation of the line is

$$\frac{x}{3} + \frac{y}{3} = 1$$

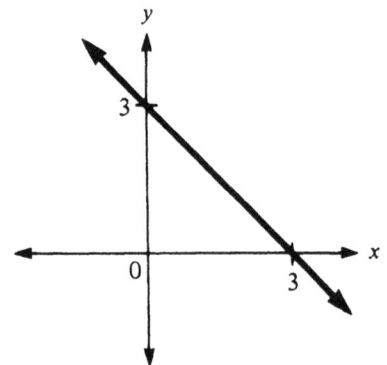

None of the choices are exactly like this equation. Multiply both sides by 3 to obtain

$$3\left(\frac{x}{3} + \frac{y}{3}\right) = 3 \times 1 \qquad \text{Use the distributive laws on the left side.}$$

$$3 \cdot \frac{x}{3} + 3 \cdot \frac{y}{3} = 3$$

$$x + y = 3$$

The correct choice is (C).    ◄

**Example 4** ▶ An equation for the line at the right is

(A) $x = 4$        (B) $y = -2$

(C) $y = 2x - 2$        (D) $y = \frac{1}{2}x - 2$

(E) $y = 4x - 2$

***Solution.*** The $x$-intercept is 4 and the $y$-intercept is $-2$. Thus the intercept form of the line is

$$\frac{x}{4} + \frac{y}{-2} = 1$$

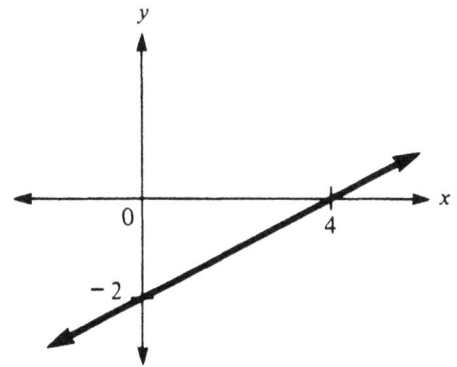

Multiply both sides by 4, the *lcd* on the left side.

$$4\left(\frac{x}{4} + \frac{y}{-2}\right) = 4 \times 1 \qquad \text{Use the distributive laws on the left side.}$$

$$4 \cdot \frac{x}{4} + 4 \cdot \frac{y}{-2} = 4$$

$$x - 2y = 4$$

Isolate $y$ on the left side, as in choices (C), (D), and (E).

$$-2y = 4 - x$$

$$\frac{-2y}{-2} = \frac{4 - x}{-2} \qquad \text{Rearrange the terms on the right side.}$$

$$y = -2 + \frac{x}{2}$$

$$y = \frac{1}{2}x - 2$$

The correct choice is (D).    ◄

**Example 5** ▶ An equation for the line at the right is

(A) $y = -2$        (B) $y = 2$

(C) $x = -2$        (D) $x + y = -2$

(E) $y = x - 2$

*Solution.* This is the *horizontal* line that passes through the point $(0, -2)$. Thus every point on the line has $y$-coordinate $-2$. The equation of the line is $y = -2$. [Choice (A)]

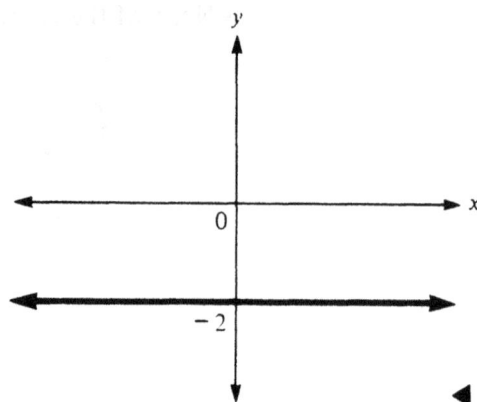

◀

**Example 6** ▶ An equation for the line at the right is

(A) $x + 5 = 0$        (B) $x - 5 = 0$

(C) $x + y = 5$        (D) $y = 5$

(E) $y = x + 5$

*Solution.* This is the *vertical* line that passes through the point $(5, 0)$. Thus every point on the line has $x$-coordinate $5$. The equation of the line is

$$x = 5$$

Add $-5$ to both sides to obtain

$$x - 5 = 5 - 5$$
$$x - 5 = 0$$

The correct choice is (B).

◀

*Try the exercises for Topic B on page 230.*

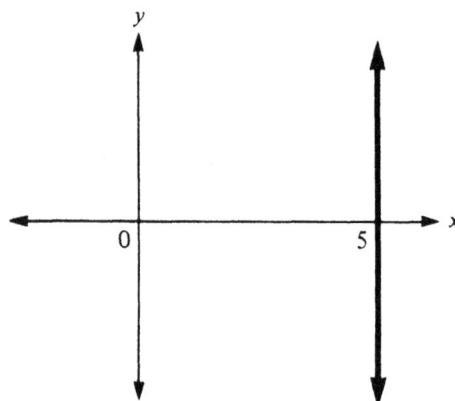

## EXERCISES

**A.** *When several choices are given, draw a circle around the correct letter.*

1. At what point does the graph of $y = x + 4$ intersect the $y$-axis?

   (A) $(4, 0)$        (B) $(0, 4)$        (C) $(0, 0)$        (D) $(-4, 0)$        (E) $(4, 4)$

2. At what point does the graph of $y = x - 2$ intersect the $y$-axis?

   (A) $(0, 2)$        (B) $(2, 0)$        (C) $(0, -2)$        (D) $(-2, 0)$        (E) $(2, -2)$

3. At what point does the graph of $y = 2x$ intersect the $y$-axis?

   (A) $(1, 2)$        (B) $(2, 2)$        (C) $(0, 2)$        (D) $(0, 0)$        (E) $(2, 0)$

4.  At what point does the graph of $y = 3x + 2$ intersect the $y$-axis?

    (A) $(2, 0)$    (B) $(0, 2)$    (C) $(3, 2)$    (D) $\left(-\frac{2}{3}, 0\right)$    (E) $(0, -2)$

5.  At what point does the graph of $y = 4x - 5$ intersect the $y$-axis?

    (A) $\left(\frac{5}{4}, 0\right)$    (B) $\left(0, \frac{5}{4}\right)$    (C) $(4, 5)$    (D) $(5, 4)$    (E) $(0, -5)$

6.  At what point does the graph of $y = \frac{1}{2}x + 1$ intersect the $y$-axis?

    (A) $\left(\frac{1}{2}, 1\right)$    (B) $(2, -1)$    (C) $(0, 1)$    (D) $\left(\frac{1}{2}, 0\right)$    (E) $\left(0, \frac{1}{2}\right)$

7.  At what point does the graph of $y = 7x - 5$ intersect the $y$-axis?
    *Answer:*

8.  At what point does the graph of $x = y + 1$ cross the $y$-axis?
    *Answer:*

9.  At what point does the graph of $x = y - 6$ cross the $y$-axis?
    *Answer:*

10. At what point does the graph of $x + y = 8$ cross the $y$-axis?
    *Answer:*

11. At what point does the graph of $x + y = 6$ cross the $x$-axis?
    *Answer:*

12. At what point does the graph of $y = 4$ cross the $y$-axis?
    *Answer:*

13. At what point does the graph of $y + 3 = 0$ cross the $y$-axis?
    *Answer:*

14. At what point does the graph of $x = 5$ cross the $x$-axis?
    *Answer:*

15. If the point $(0, b)$ lies on the graph of $y = x + 6$, then $b =$

    (A) 6    (B) $-6$    (C) 3    (D) 1    (E) 0

16. If the point $(0, b)$ lies on the graph of $y = x - 3$, then $b =$

    (A) 3    (B) $\frac{1}{3}$    (C) $-3$    (D) 0    (E) $-\frac{1}{3}$

17. If the point $(0, b)$ lies on the graph of $y = 2x - 8$, then $b =$

    (A) 8    (B) $-8$    (C) 2    (D) 4    (E) $-4$

18. If the point $(0, b)$ lies on the graph of $y = 3x + 4$, then $b =$

    (A) 3    (B) 4    (C) $-4$    (D) $\frac{4}{3}$    (E) $-\frac{4}{3}$

19. If the point $(0, b)$ lies on the graph of $y = 2x - 5$, then $b =$

    (A) $\frac{5}{2}$    (B) $-5$    (C) $\frac{2}{5}$    (D) 2    (E) 5

20. If the point $(0, b)$ lies on the graph of $4x + y = 0$, then $b =$

    (A) 4          (B) $-4$          (C) $\frac{1}{4}$          (D) $-\frac{1}{4}$          (E) 0

21. If the point $(0, b)$ lies on the graph of $x + y = 6$, then $b =$

22. If the point $(0, b)$ lies on the graph of $2x + 3y = 6$, then $b =$

23. If the point $(0, b)$ lies on the graph of $5y = x + 10$, then $b =$

24. If the point $(0, b)$ lies on the graph of $2x + 3y = 12$, then $b =$

25. If the point $(0, b)$ lies on the graph of $4y + 2x = 16$, then $b =$

26. If the point $(0, b)$ lies on the graph of $3x - 2y = 10$, then $b =$

27. If the point $(0, b)$ lies on the graph of $y = 7$, then $b =$

28. If the point $(0, b)$ lies on the graph of $3x = 4y$, then $b =$

B. *Find an equation for the given line. Draw a circle around the correct letter.*

29. An equation for the line
    below is

    (A) $y = x + 3$          (B) $y = x - 3$

    (C) $y = 3$              (D) $x + y = 3$

    (E) $x - y = 3$

30. An equation for the line
    below is

    (A) $x + y = 6$          (B) $x + y = 2$

    (C) $y = 2x + 6$        (D) $y = 6 - 3x$

    (E) $y = 6 + 3x$

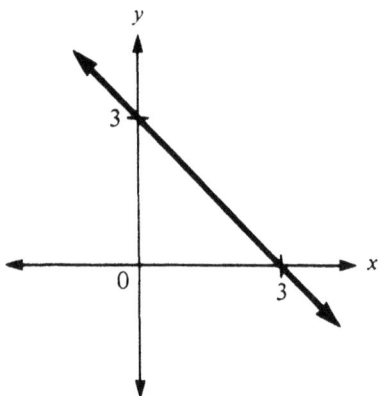

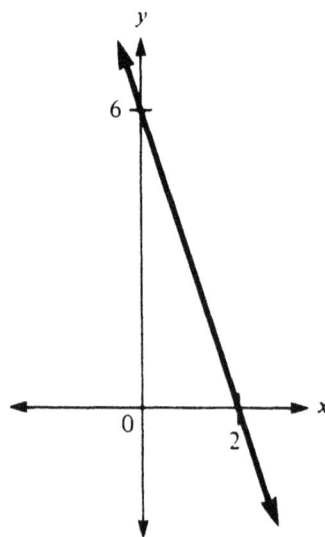

31. An equation for the line below is

   (A) $y = 4$          (B) $x + y = 4$

   (C) $x = 4$          (D) $x - y = 4$

   (E) $y = x + 4$

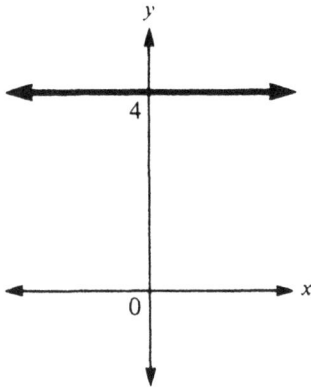

32. And equation for the line below is

   (A) $x - 3 = 0$      (B) $x + 3 = 0$

   (C) $y = 3$          (D) $y = -3$

   (E) $y = x - 3$

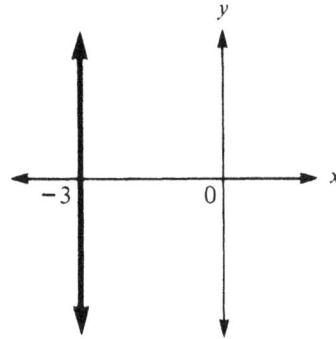

33. An equation for the line at the right is

   (A) $x = 2$

   (B) $y = -2$

   (C) $x + y = 2$

   (D) $y = x + 2$

   (E) $y = x - 2$

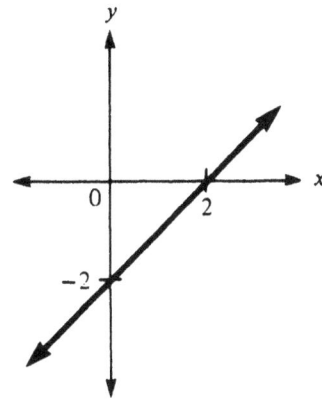

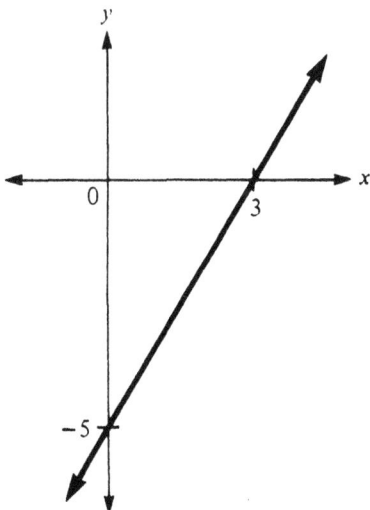

34. An equation for the line at the left is

   (A) $3x - 5y = 0$

   (B) $5x - 3y = 15$

   (C) $x = 3$

   (D) $5y - 3x = 15$

   (E) $3y - 5x = 15$

**35.** An equation for the line below is

(A) $2y - 4x = 0$    (B) $y = -4x + 2$

(C) $x + y = 2$    (D) $2y - x = 4$

(E) $2x - y = 1$

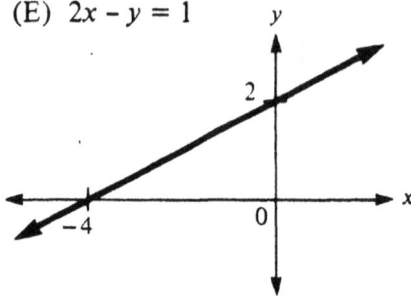

**36.** An equation for the line below is

(A) $x + y = -1$    (B) $y = -x$

(C) $x + y - 1 = 0$    (D) $x = -1$

(E) $y = -1$

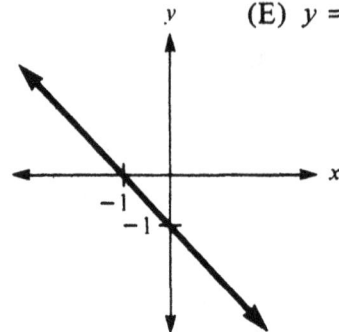

**37.** An equation for the line at the right is

(A) $x = 5$    (B) $y = -1$

(C) $x + y = 5$    (D) $5y = x - 1$

(E) $5y = x - 5$

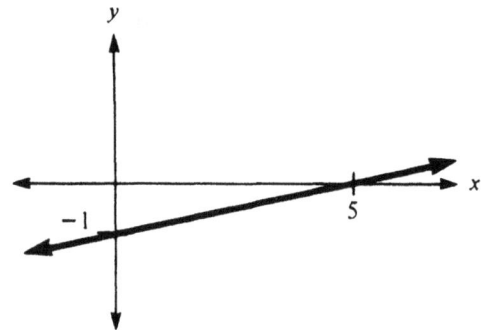

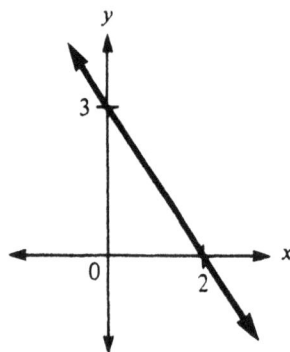

**38.** An equation for the line at the left is

(A) $3y = 2x$    (B) $2y = 3x$

(C) $y = \frac{3}{2}x + 3$    (D) $\frac{y}{3} + \frac{x}{2} = 1$

(E) $\frac{y}{2} + \frac{x}{3} = 1$

**39.** An equation for the line below is

(A) $x + y = 2$    (B) $x - y = 2$

(C) $y - x = 2$    (D) $y = 2$

(E) $y = 2x$

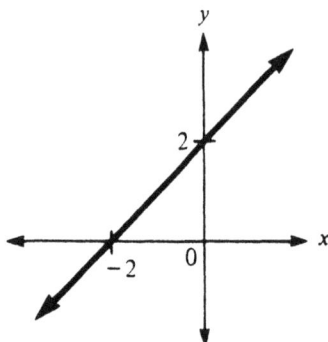

**40.** An equation for the line below is

(A) $x + y = 4$

(B) $y = x + 4$

(C) $y - 4x = 1$

(D) $y - 4x = 4$

(E) $x - 4y = 4$

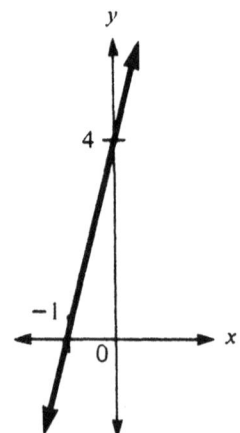

# 33.  Systems of Equations

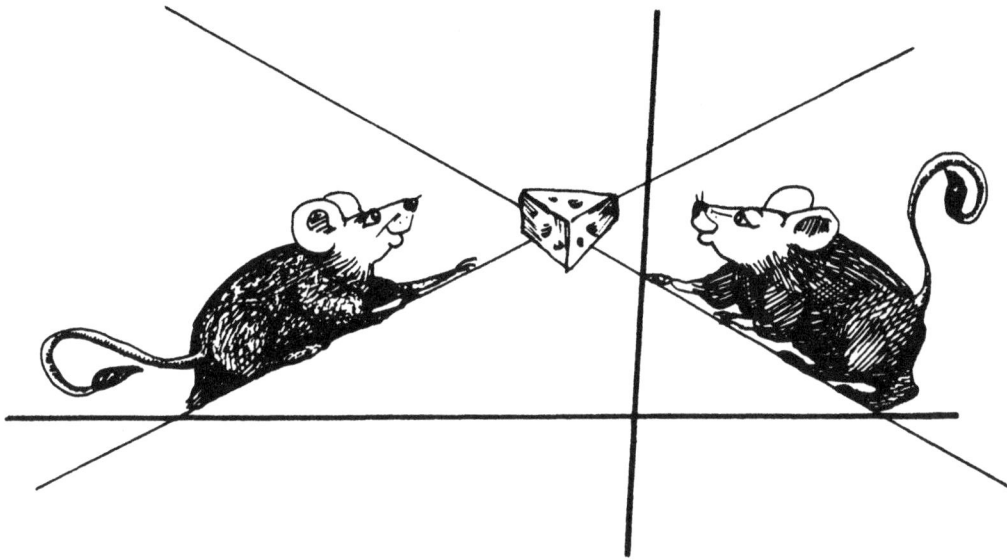

Here we will consider how to solve a system of two equations in two variables. A knowledge of systems of equations is helpful in solving certain verbal problems.

## A. SOLUTION OF A SYSTEM

DEFINITION

> Two equations, taken together, form a **system of equations.**

For example, the equations

$$x + y = 9$$
$$2x - y = 6$$

together form a system of equations.

Recall that the point $(a, b)$ is called a *solution* of an equation in the variables $x$ and $y$ if a true statement results when $a$ is substituted for $x$ and $b$ for $y$.

Thus the point $(5, 4)$ is a solution of the equation

$$x + y = 9$$

because when we substitute 5 for $x$ and 4 for $y$, we obtain

$$5 + 4 = 9 \qquad true$$

Furthermore, $(6, 3)$ is another solution of $x + y = 9$ because

$$6 + 3 = 9 \qquad true$$

Observe that $(5, 4)$, the first point considered, is also a solution of the equation

$$2x - y = 6$$

because when we substitute 5 for $x$ and 4 for $y$ in this equation, we obtain

$$2(5) - 4 = 6$$
$$10 - 4 = 6 \qquad \textit{true}$$

However, $(6, 3)$, the second point considered, is not a solution of $2x - y = 6$ because

$$2(6) - 3 = 6$$
$$12 - 3 = 6 \qquad \textit{false}$$

We will say that $(5, 4)$ is a *solution of the system of equations*

$$x + y = 9$$
$$2x - y = 6$$

but that $(6, 3)$ is not.

DEFINITION

> The point $(a, b)$ is a **solution of a system of equations** if $(a, b)$ is a solution of *both* equations of the system.

Before learning how to *solve* a system of equations, we will first consider how to *check* whether a given point $(a, b)$ is, indeed, a solution of a system. By definition, the point must be a solution of *both* equations of the system. Thus substitute $a$ for $x$ and $b$ for $y$ in each of the two given equations. If you find that the point is a solution of the first equation, then check it also in the second equation. But if the given point is *not* a solution of the first equation, it cannot be a solution of the system. It is therefore *not* necessary to check the second equation.

**Example 1 ▶**    Is $(2, 5)$ a solution of the system

$$2x + y = 9$$
$$3x - y = 1$$

*Solution.*  First substitute 2 for $x$ and 5 for $y$ in the equation $2x + y = 9$.

$$2(2) + 5 = 9$$

$$9 = 9 \qquad \textit{true}$$

Thus $(2, 5)$ is a solution of the first equation. Now substitute 2 for $x$ and 5 for $y$ in the second equation, $3x - y = 1$.

$$3(2) - 5 = 1$$
$$6 - 5 = 1 \qquad true$$

Because $(2, 5)$ is also a solution of the second equation, it is a solution of the given *system*. ◄

**Example 2 ▶** Is $(2, 2)$ a solution of the system

$$2x + 3y = 10$$
$$x + y = \phantom{0}6$$

*Solution.* Substitute 2 for $x$ and 2 for $y$ in the first equation, $2x + 3y = 10$.

$$2(2) + 3(2) = 10$$
$$4 + 6 = 10 \qquad true$$

Thus $(2, 2)$ is a solution of the first equation. Now substitute 2 for $x$ and 2 for $y$ in the second equation, $x + y = 6$.

$$2 + 2 = 6 \qquad false$$

Because $(2, 2)$ is *not* a solution of the second equation, it is *not* a solution of the system.

In each equation we will be considering, $x$ and $y$ appear only to the first power. Thus the graph of each such equation is a straight line. *A solution of a system of such equations represents the intersections of the lines whose equations comprise the system.* In fact, two (different) intersecting lines meet at a single point. The corresponding system of equations has a single solution. For the system

$$2x + y = 9$$
$$3x - y = 1$$

of Example 1, the lines with these equations are pictured at the right. Note that the intersection of these lines is the point $(2, 5)$, which we found to be the solution of the system.

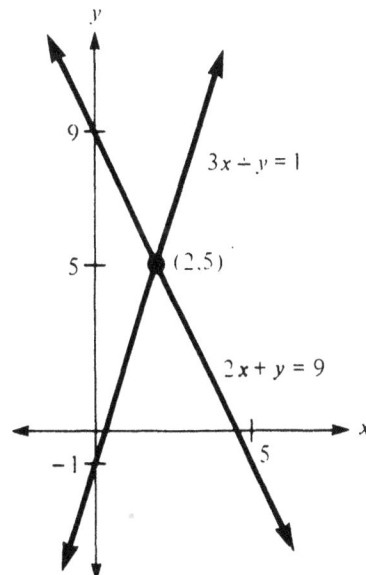

Although we will *not* study them here, there are systems, such as

$$x + y = 1$$
$$x + y = 5$$

in which *the equations represent parallel lines.* Such a system has *no* solution because there is no point that lies on *both* lines. (See the figure at the right.) There are also systems, such as

$$x + y = 3$$
$$2x + 2y = 6$$

in which *both equations represent the same line.* Every point on this line is then a solution of the system. Note that here, the second equation is obtained by multiplying both sides of the first equation by 2. (See the figure at the right.)

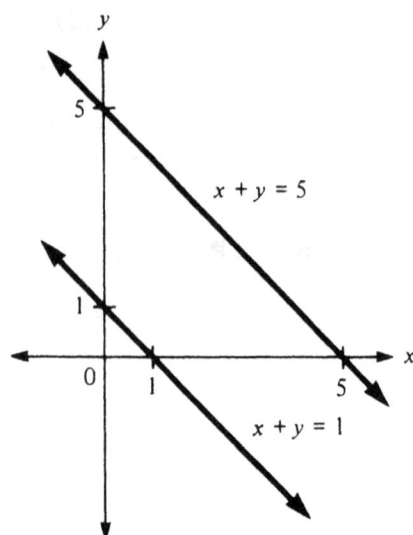

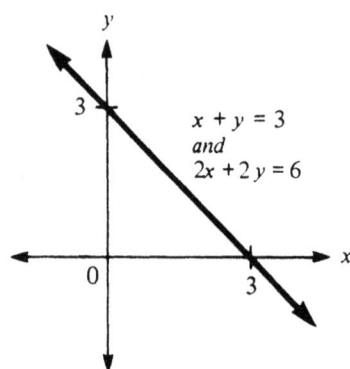

*Try the exercises for Topic A on page 239.*

## B. SOLVING A SYSTEM BY ADDING

We can solve a system of equations by first adding the corresponding sides of the equations to eliminate one of the variables.

**Example 3 ▶**    Solve the system:

$$x + y = 10$$
$$x - y = 4$$

*Solution.* Observe that $y$ appears in the first equation with coefficient 1 and in the second equation with coefficient $-1$. We can add the corresponding sides to eliminate $y$.

$$
\begin{array}{rl}
x + y & = 10 \\
\underline{x - y} & \underline{= 4} \\
2 \quad\;\; & = 14 \qquad \text{Divide both sides by 2.} \\
x \quad\; & = 7
\end{array}
$$

Now substitute 7 for $x$ in either equation to find $y$. We will use the first equation.

$$7 + y = 10 \qquad \text{Subtract 7 from both sides.}$$
$$y = 3$$

Thus $x = 7$ and $y = 3$. The solution of the system is the point $(7, 3)$. This point is the intersection of the lines with the given equations. See the figure at the right.

We can also *check* that $(7, 3)$ is a solution by substituting 7 for $x$ and 3 for $y$ in each of the given equations. Thus

$$7 + 3 = 10 \quad \textit{true}$$

$$7 - 3 = 4 \quad \textit{true}$$

The check confirms that $(7, 3)$ is a solution.

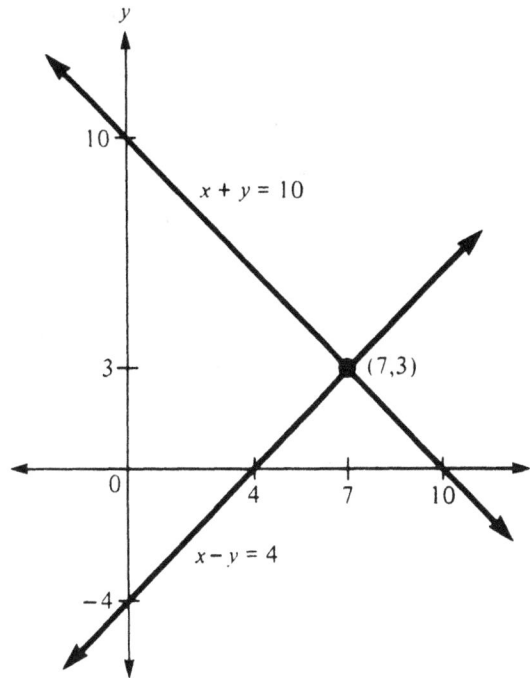

**Example 4 ▶**  If $2x + y = 5$ and $2x + 5y = 1$, then

(A) $x = 1$ and $y = 3$   (B) $x = 2$ and $y = 1$   (C) $x = 3$ and $y = -1$

(D) $x = 0$ and $y = 5$   (E) $x = 0$ and $y = \dfrac{1}{5}$

***Solution.***  Observe that $x$ appears in *both* equations with the same coefficient, 2. To eliminate $x$, *subtract* the corresponding sides of the first equation from those of the second equation.

$$\begin{array}{rl} 2x + 5y = & 1 \\ \underline{2x + \phantom{5}y = \phantom{-}5} & \text{Subtract.} \\ 4y = -4 & \text{Divide both sides by 4.} \\ y = -1 & \end{array}$$

Now substitute $-1$ for $y$ in the first equation, $2x + y = 5$, to find $x$.

$$2x + (-1) = 5 \quad \text{Subtract } -1 \text{ from, or add 1 to, both sides.}$$

$$2x = 6$$

$$x = 3$$

The solution is $(3, -1)$. See the figure below. Thus $x = 3$ and $y = -1$ [choice (C)].

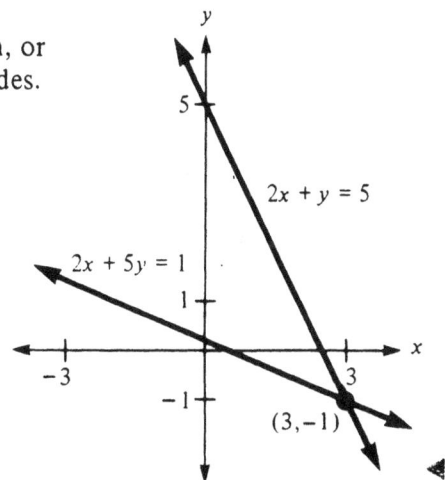

Sometimes, to solve a system of equations, it is best to multiply both sides of one of the equations by a number before adding or subtracting.[1]

[1] The case in which both equations must be modified will not be considered here.

**Example 5 ▶**    The graphs of $x + y = 5$ and $3x - 2y = 5$ meet at the point where

(A) $x = 0, y = 5$          (B) $x = 5, y = 0$          (C) $x = 1, y = 4$

(D) $x = 2, y = 3$          (E) $x = 3, y = 2$

*Solution.* If we multiply both sides of the first equation by 2, we obtain an equivalent equation (that is, an equation with the same solutions),

$$2x + 2y = 10$$

Now $y$ appears with coefficient 2 in this equation and with coefficient $-2$ in the given second equation. Add these equations to eliminate $y$.

$$\begin{array}{rl} 2x + 2y &= 10 \\ 3x - 2y &= \ \ 5 \\ \hline 5x \quad\ \ &= 15 \qquad \text{Divide both sides by 5.} \\ x \quad\ \ &= \ \ 3 \end{array}$$

Substitute 3 for $x$ in the given first equation, $x + y = 5$, to find $y$.

$$3 + y = 5 \qquad \text{Subtract 3 from both sides.}$$

$$y = 2$$

Thus the point $(3, 2)$ is the solution of the given system of equations. Therefore the lines with equations $x + y = 5$ and $3x - 2y = 5$ meet at the point where $x = 3$ and $y = 2$ [choice (E)]. See the figure at the right.

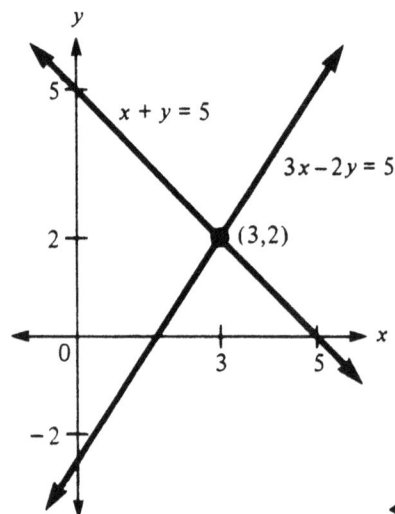

*Try the exercises for Topic B on page 240.*

## C. SOLVING A SYSTEM BY SUBSTITUTING

Sometimes, the equations of a system are presented in a form that enables you, easily, to substitute for one variable in terms of the other. This method will be used in solving applied problems in Section 34.

**Example 6 ▶**    If $y = 2x + 3$  and  $3x + 2y = 20$, then

(A) $x = 0, y = 3$          (B) $x = 1, y = 5$          (C) $x = 4, y = 4$

(D) $x = 2, y = 7$          (E) $x = 0, y = 10$

*Solution.* According to the first equation,

$$y = 2x + 3$$

Substitute $2x + 3$ for $y$ in the second equation,

$$3x + 2y = 20$$

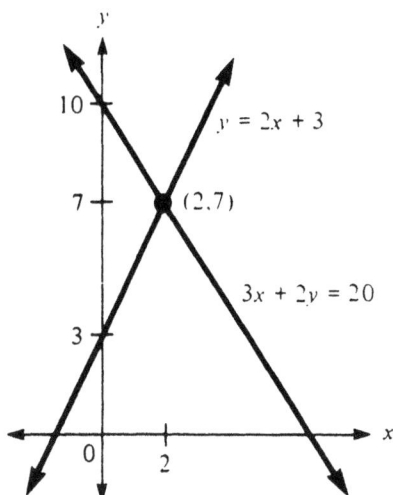

to obtain

$$3x + 2(2x + 3) = 20 \qquad \text{Simplify the left side.}$$
$$3x + 4x + 6 = 20$$
$$7x + 6 = 20 \qquad \text{Subtract 6 from both sides.}$$
$$7x = 14$$
$$x = 2$$

Now substitute 2 for $x$ in the first equation, $y = 2x + 3$.

$$y = 2(2) + 3$$
$$= 4 + 3$$
$$= 7$$

Thus $x = 2$ and $y = 7$ [choice (D)]. See the figure. ◄

*Try the exercises for Topic C on page 241.*

## EXERCISES

**A.** *Determine whether the given point is the solution of the system. Answer "yes" or "no."*

1. Is (8, 4) a solution of the system

$$x + y = 12$$
$$x - y = 4$$

*Answer:*

2. Is (3, 4) a solution of the system

$$2x + y = 10$$
$$x + 2y = 11$$

*Answer:*

3. Is (2, 2) a solution of the system

$$5x + y = 12$$
$$3x - y = -4$$

*Answer:*

4. Is (5, 2) a solution of the system

$$2x - 5y = 0$$
$$x + 2y = 9$$

*Answer:*

5. Is (7, 1) a solution of the system

$$3x - 11y = 10$$
$$4x - 20y = 8$$

*Answer:*

6. Is (2, –3) a solution of the system

$$4x + y = 5$$
$$x - y = -1$$

*Answer:*

7. Is (4, 0) a solution of the system

$$2x + 3y = 8$$
$$3x - 5y = 12$$

*Answer:*

8. Is (1, –1) a solution of the system

$$5x + 3y = 2$$
$$4x - 4y = 0$$

*Answer:*

9. Is (6, 3) a solution of the system

$$2x - 3y = 3$$
$$3x - 4y = 6$$

*Answer:*

10. Is (0, –1) a solution of the system

$$7x - 3y = 3$$
$$5x + 4y = -4$$

*Answer:*

**B.** *In Exercises 11–18 find the solution of each system.*

11.  $x + y = 6$
     $x - y = 2$

     *Answer:*

12.  $x + y = 9$
     $x - y = 5$

     *Answer:*

13.  $x + y = 10$
     $2x - y = 2$

     *Answer:*

14.  $x + y = 7$
     $3x + y = 9$

     *Answer:*

15.  $2x + y = 9$
     $2x + 3y = 11$

     *Answer:*

16.  $2x + 3y = 6$
     $4x - 3y = 12$

     *Answer:*

17.  $5x - 2y = 1$
     $2x + 4y = 10$

     *Answer:*

18.  $4x - 2y = 10$
     $3x - 4y = 5$

     *Answer:*

*When several choices are given, draw a circle around the correct letter.*

19.  If $x + y = 12$ and $x - y = 2$, then

     (A) $x = 6, y = 6$          (B) $x = 8, y = 4$          (C) $x = 9, y = 3$

     (D) $x = 5, y = 7$          (E) $x = 7, y = 5$

20.  If $x + y = 6$ and $x - y = 10$, then

     (A) $x = 4, y = 2$          (B) $x = 6, y = 0$          (C) $x = 0, y = 6$

     (D) $x = 8, y = -2$         (E) $x = 12, y = 2$

21.  If $x + y = 10$ and $2x - y = -1$, then

     (A) $x = 7, y = 3$          (B) $x = 3, y = 7$          (C) $x = 5, y = 5$

     (D) $x = 9, y = 1$          (E) $x = 4, y = 6$

22.  If $x + y = 20$ and $x - y = 4$, then

     (A) $x = 10, y = 10$        (B) $x = 16, y = 4$         (C) $x = 9, y = 11$

     (D) $x = 12, y = 8$         (E) $x = 8, y = 12$

23.  If $x + 2y = 6$ and $x - 3y = 1$, then

     (A) $x = 4, y = 1$          (B) $x = 0, y = 3$          (C) $x = 6, y = 0$

     (D) $x = 8, y = -1$         (E) $x = 2, y = 2$

24.  If $2x + y = 9$ and $3x - y = 6$, then

(A) $x = 1, y = 7$    (B) $x = 2, y = 5$    (C) $x = 3, y = 3$

(D) $x = 2, y = 0$    (E) $x = 0, y = 9$

25.  Solve the system
$$2x - y = 9$$
$$x - y = 4$$
(A) $(6, 3)$    (B) $(5, 1)$    (C) $(8, 7)$    (D) $(0, -9)$    (E) $(9, 5)$

26.  Solve the system
$$2x + y = 8$$
$$x + 3y = 9$$
(A) $(0, 8)$    (B) $(4, 0)$    (C) $(1, 6)$    (D) $(2, 4)$    (E) $(3, 2)$

27.  Solve the system
$$4x + 3y = 1$$
$$2x - 4y = 6$$
*Answer:*

28.  Solve the system
$$2x + 3y = 8$$
$$x + 6y = 13$$
*Answer:*

29.  The graphs of $x + y = 5$ and $2x - y = 1$ meet at the point where

(A) $x = 2, y = 3$    (B) $x = 4, y = 1$    (C) $x = 5, y = 0$

(D) $x = 0, y = 5$    (E) $x = 6, y = -1$

30.  The graphs of $3x + y = 10$ and $2x + 3y = 9$ meet at the point where

(A) $x = 0, y = 10$    (B) $x = 4, y = -2$    (C) $x = 3, y = 1$

(D) $x = 2, y = 4$    (E) $x = 5, y = -5$

31.  The graphs of $2x + 5y = 9$ and $4x - 3y = 5$ meet at the point where
$x =$          and $y =$

32.  The graphs of $4x + 3y = 2$ and $x + y = 0$ meet at the point where
$x =$          and $y =$

C.  *When several choices are given, draw a circle around the correct letter.*

33.  If $y = x + 3$ and $3x - y = 9$, then

(A) $x = 0, y = 3$    (B) $x = 1, y = 4$    (C) $x = 2, y = 5$

(D) $x = 6, y = 9$    (E) $x = 3, y = 0$

**34.** If $y = 2x - 4$ and $3x - 2y = 3$, then

    (A) $x = 2, y = 0$          (B) $x = 0, y = -2$          (C) $x = 5, y = 6$

    (D) $x = 1, y = 0$          (E) $x = 8, y = 12$

**35.** If $y = 7 - x$ and $3x - 4y = 0$, then

    (A) $x = 0, y = 0$          (B) $x = 7, y = 0$          (C) $x = 0, y = 7$

    (D) $x = 4, y = 3$          (E) $x = -4, y = 3$

**36.** If $y = 5 - 4x$ and $4x - 3y = 1$, then

    (A) $x = 0, y = 5$          (B) $x = 1, y = 1$          (C) $x = 2, y = -3$

    (D) $x = -1, y = 9$          (E) $x = -2, y = 13$

**37.** If $x = 3 - y$ and $5x - 2y = 1$, then    $x =$          and $y =$

**38.** If $x = 10 - 2y$ and $3y - x = 0$, then    $x =$          and $y =$

**39.** If $x = 9 - 2y$ and $y = x - 3$, then    $x =$          and $y =$

**40.** If $y = 4x$ and $3x - y + 4 = 0$, then    $x =$          and $y =$

# 34.  Applied Problems

Some percent problems in Section 14 were stated verbally. Now we consider other types of verbal problems. You must formulate the problem in terms of an equation, and then solve the equation in order to solve the given problem.

## A. EQUATIONS IN ONE VARIABLE

**Example 1** ▶ A drama group charges $8 a ticket to see its production of *Hamlet*. If its expenses are $1200, how many tickets must it sell to make a profit of $800?

(A) 100    (B) 125    (C) 150    (D) 200    (E) 250

*Solution.*  Let $x$ be the number of tickets the group must sell. Then $8x$ is the income (in dollars) from the sale of these tickets. Use the formula

Income − Expenses = Profit

to set up the equation.

$$8x - 1200 = 800 \qquad \text{Add 1200 to both sides.}$$
$$8x = 2000$$
$$x = \frac{2000}{8}$$
$$x = 250$$

Thus 250 tickets must be sold [choice (E)].    ◀

**Example 2** ▶ A rope 16 feet long is cut into two pieces. If one piece is three times as long as the other, find the length of the shorter piece.

*Solution.* Let $x$ be the length (in feet) of the shorter piece. The longer piece is three times as long. Thus its length is expressed by $3x$.

$$\underbrace{\text{Length of shorter piece}}_{x} + \underbrace{\text{length of longer piece}}_{3x} = 16 \text{ feet}$$

$$x + 3x = 16$$
$$4x = 16$$
$$x = 4$$

The length of the shorter piece is 4 feet.    ◀

*Try the exercises for Topic A on page 248.*

## B. PROPORTIONS

Recall that a *proportion* is an equation of the form

$$\frac{a}{b} = \frac{c}{d}$$

Proportions were considered in Section 28. Many applied problems are solved by setting up proportions.

**Example 3 ▶**   Three out of every 5 students who take freshman psychology pass the final. If 120 students pass the final, how many take the course?

(A) 24          (B) 72          (C) 200          (D) 360          (E) 600

*Solution.* Let $x$ be the number of students who take freshman psychology. We are told that 3 *out of* 5 students pass the course. In other words, $\frac{3}{5}$ of the students pass the course. Thus we can set up the proportion

$$\frac{\text{number who pass the final}}{\text{number who take freshman psych}} = \frac{3}{5}$$

Because 120 students pass the final and $x$ students take freshman psych, this proportion can be written as

$$\frac{120}{x} = \frac{3}{5} \qquad \text{Cross-multiply.}$$
$$120 \times 5 = 3x$$
$$600 = 3x$$
$$200 = x$$

The correct choice is (C).    ◀

**Example 4 ▶**   25% of the 80 students at a school of music are women. The school decides to admit just enough additional women so that 50% of the students will then be women. How many additional women must be admitted?

***Solution.*** Here the first sentence tells us how many women are presently at the music school. Recall that "of" in a sentence like this indicates multiplication.

$$\underbrace{25\%}_{.25} \ \overset{\vdots}{\times} \ \underbrace{\text{of the 80 students}}_{80} \ \overset{\vdots}{=} \ \text{(number of) women}$$

$$
\begin{array}{ll}
.25 & \longleftarrow \text{2 decimal digits} \\
\underline{\times\ 80} & \longleftarrow + \text{0 decimal digits} \\
20.00 & \phantom{\longleftarrow}\ \text{2 decimal digits}
\end{array}
$$

Thus presently, there are 20 women.

Now let $x$ be the number of *additional* women that must be admitted. (No additional men are to be admitted.) Thus the *totals* for women and for *all* students are as follows:

|  | Number at Present | + | Additional Number |
|---|---|---|---|
| Women | 20 | + | $x$ |
| All students | 80 | + | $x$ |

Because 50% of the students will then be women and because $50\% = \frac{1}{2}$, set up the proportion

$$\frac{\text{total number of women}}{\text{total number of students}} = \frac{1}{2}$$

$$\frac{20 + x}{80 + x} = \frac{1}{2} \qquad \text{Cross-multiply.}$$

$$2(20 + x) = 1(80 + x) \qquad \text{Use the distributive laws.}$$

$$40 + 2x = 80 + x$$

$$40 + 2x - x = 80 + x - x$$

$$40 + x = 80$$

$$x = 40 \qquad\qquad\qquad\qquad\blacktriangleleft$$

*Try the exercises for Topic B on page 249.*

## C. EQUATIONS IN TWO VARIABLES

Often, it is convenient to formulate a problem in terms of a system of equations in two variables.

**Example 5** ▶ Among 95 passengers on a flight to Boston, one less than twice the number of 1st-class passengers are flying coach. How many are flying coach?

***Solution.*** Let $x$ be the number of 1st-class passengers and let $y$ be the number of coach passengers. Then from the first sentence,

$$x + y = 95$$

From the second sentence,

One less than twice the number of 1st-class passengers are flying coach.

$$-1 + \qquad\qquad 2x \qquad\qquad\quad \doteq \quad y$$

Use the second equation to substitute $-1 + 2x$ for $y$ in the first equation, $x + y = 95$.

| | |
|---|---|
| $x + (-1 + 2x) = 95$ | Combine like terms on the left side. |
| $3x - 1 = 95$ | Add 1 to both sides. |
| $3x = 96$ | |
| $x = 32$ | |

Now use the first equation, $x + y = 95$, to find $y$.

$$32 + y = 95$$
$$y = 63$$

Thus 63 passengers are flying coach.    ◀

**Example 6 ▶**    On a math exam Susan's grade was 8 points higher than Teresa's grade. The *average* of their grades was 83. Find Susan's grade.

(A) 75    (B) 79    (C) 83    (D) 87    (E) 91

*Solution.* Let $s =$ Susan's grade and let $t =$ Teresa's grade. Then, according to the first sentence,

Susan's grade was 8 points higher than Teresa's grade.

$$s \qquad \doteq \qquad 8 + \qquad\qquad t$$

Recall that the average of *two* numbers (Section 3), is their sum divided by 2. Here, the average of their grades is

$$\frac{s + t}{2}$$

According to the second sentence,

$$\frac{s + t}{2} = 83$$

Use the first equation to substitute $8 + t$ for $s$ in the second equation.

| | |
|---|---|
| $\dfrac{8 + t + t}{2} = 83$ | Multiply both sides by 2. |
| $8 + t + t = 166$ | Combine like terms on the left side. |
| $8 + 2t = 166$ | Subtract 8 from both sides. |
| $2t = 158$ | |
| $t = 79$ | |

To find $s$, which represents Susan's score, use the first equation, $s = 8 + t$.

$$s = 8 + 79$$

$$s = 87$$

The correct choice is (D).                                    ◀

*Try the exercises for Topic C on page 250.*

## D. PROPORTIONS IN TWO VARIABLES

Occasionally, a problem is most easily solved by setting up two proportions in two variables, as illustrated in Example 7.

**Example 7** ▶ At a party, 70% of the people are women. When twenty women leave (and all of the men remain), 50% of the people are then women. How many men are at the party?

(A) 10          (B) 12          (C) 15          (D) 20          (E) 30

*Solution.* Use

$$70\% = \frac{7}{10} \qquad \text{and} \qquad 50\% = \frac{1}{2}$$

Let $x$ be the number of women *originally* at the party and let $y$ be the number of men there. Then $x + y$ is the number of people, men or women, originally at the party. Because 70% are women,

$$\frac{\text{number of women (originally)}}{\text{number of people (originally)}} = 70\%$$

$$\frac{x}{x + y} = \frac{7}{10} \qquad\qquad \text{Cross-multiply.}$$

$$10x = 7(x + y) \qquad \text{Use the distributive laws on the right side.}$$

$$10x = 7x + 7y \qquad \text{Subtract } 7x \text{ from both sides.}$$

(*)    $$3x = 7y$$

Now use the second sentence of the problem. When twenty women leave, the number of women at the party is $x - 20$. All of the men remain, so that the number of people present is $(x - 20) + y$. Because 50% are then women,

$$\frac{\text{number of women (afterward)}}{\text{number of people (afterward)}} = 50\%$$

$$\frac{x - 20}{x - 20 + y} = \frac{1}{2} \qquad\qquad \text{Cross-multiply.}$$

$$2(x - 20) = 1(x - 20 + y)$$

$$2x - 40 = x - 20 + y$$

$$2x - 40 + 40 = x - 20 + 40 + y$$

$$2x = x + 20 + y \qquad \text{Subtract } x \text{ from both sides.}$$

$$x = 20 + y$$

Now substitute $20 + y$ for $x$ in equation (*), $3x = 7y$. Thus

$3(20 + y) = 7y$    Use the distributive law on the left side.

$60 + 3y = 7y$

$60 = 4y$

$15 = y$

Thus the number of men present is 15, choice (C). To *check* that this is correct, use $x = 20 + y$, so that

$x = 20 + 15 = 35$

There were 35 women originally at the party. Now consider the given information,

$$\frac{\text{number of women (originally)}}{\text{number of people (originally)}} = \frac{35}{50} = \frac{7}{10} = 70\%$$

$$\frac{\text{number of women (afterward)}}{\text{number of people (afterward)}} = \frac{35 - 20}{50 - 20}$$

$$= \frac{15}{30}$$

$$= \frac{1}{2}$$

$$= 50\%$$

Both pieces of information check.    ◀

*Try the exercises for Topic D on page 250.*


## EXERCISES

**A.** *In Exercises 1-4, draw a circle around the correct letter.*

1.  At a football game each ticket costs $5. If expenses for the game total $2500, how many tickets must be sold in order to make a profit of $2000?

    (A) 100        (B) 400        (C) 500        (D) 900        (E) 1000

2.  A publisher charges bookstores $9 per dictionary. If expenses amount to $4500, how many dictionaries must be sold in order for the publisher to earn $8100 in profits?

    (A) 500        (B) 900        (C) 1400        (D) 2000        (E) 9000

3.  An opera company charges $12 per ticket to its performance of *Aida*. Its expenses total $7200, and the company *loses* $2700 on this venture. How many tickets does it sell?

    (A) 225        (B) 375        (C) 600        (D) 825        (E) 975

4.  A 32-foot board is sawed into two pieces. If one piece is 5 feet longer than the other, find the length of the *longer* piece.

    (A) 27 feet                (B) 21 feet                (C) 19 feet

    (D) 18 feet 6 inches       (E) 19 feet 6 inches

5. A 24-meter cable is sawed into two pieces. The longer piece is twice as long as the shorter piece. Find the length of the shorter piece. *Answer:*

6. It costs a manufacturer $10 to produce each video game. In addition, there is a general overhead cost of $20,000. He decides to produce 5000 games. If he receives $20 per game from a wholesaler, how many games must he sell to break even? *Answer:*

7. In Exercise 6, how many games must the manufacturer sell to make a profit of $10,000? *Answer:*

8. A novelty manufacturer finds that it costs her $2 to produce an item, and that there is a general overhead cost of $500. If she is willing to spend $5000 on this venture, how many items can she produce? *Answer:*

**B.** *When several choices are given. draw a circle around the correct letter.*

9. The width and length of an 8 X 10 photograph are enlarged proportionally. If the width of the enlargement is 12 inches, what is the length?

   (A) 9 inches        (B) 12 inches        (C) 15 inches
   (D) 16 inches       (E) 20 inches

10. A saleswoman receives a $36 commission on a $200 sale. At this rate, how much does she receive on a $450 sale?

    (A) $64      (B) $72      (C) $80      (D) $81      (E) $90

11. A typist makes 4 errors on 14 pages. At this rate, how many errors will he make on 49 pages?

    (A) 10      (B) 12      (C) 14      (D) 16      (E) 18

12. One month an agency finds that 2 out of every 5 cars sold are sports cars. If the agency sells 70 sports cars that month, how many cars does it sell?

    (A) 35      (B) 90      (C) 105      (D) 175      (E) 350

13. A hostess makes 20 cups of coffee for 14 guests. At this rate, how many cups should she make for 35 guests?
    *Answer:*

14. A basketball player makes 3 out of 5 free throws. At this rate, how many free throws must she attempt in order to make 15 of them?
    *Answer:*

15. Three out of every 10 students at a college take French. If there are 660 students at the college, how many take French? *Answer:*

16. Two out of every 9 medical students drop out before graduation. In order to graduate 350 doctors, how many medical students should be admitted?

    (A) 375      (B) 400      (C) 425      (D) 450      (E) 500

17. Thirty percent of the 500 students at a college are from California. In order to have half the students from California, how many additional California students must be admitted?

    (A) 150      (B) 200      (C) 500      (D) 650      (E) 700

18. Ten percent of the 600 employees in a factory are women. In order to have 50% women employees in the factory, how many additional women must be hired? *Answer:*

19. Forty percent of the 1200 students at a college are from minority groups. In order to have 50% of the students from minority groups, how many additional minority group students must be admitted? *Answer:*

C. *In Exercises 20-22, draw a circle around the correct letter.*

20. Among 80 students in a dormitory, four less than three times the number of foreign-born students are American-born. How many of these students are foreign-born?

    (A) 19        (B) 20        (C) 21        (D) 28        (E) 29

21. Among 95 workers in a factory, the number of skilled workers is 5 more than twice the number of unskilled. How many skilled workers are there in the factory?

    (A) 30        (B) 35        (C) 55        (D) 65        (E) 70

22. In a school election 1440 votes are cast for the two candidates. One candidate wins by 10 votes. How many votes does this candidate receive?

    (A) 720       (B) 725       (C) 730       (D) 735       (E) 740

23. Jules and Jim together have $500. Jules has $20 more than Jim. How much does Jules have? *Answer:*

24. Mozart wrote four less than five times the number of symphonies that Beethoven wrote. Altogether, they wrote 50 symphonies. How many did Mozart write? *Answer:*

25. Bill's chemistry grade was 8 points higher than Joe's. The average of their grades was 75. Find Bill's grade. *Answer:*

26. Jerry's math grade was 14 points less than Harriet's. If the average of their grades was 86, what was Harriet's grade? *Answer:*

27. Ann received a grade of 88 on her history exam, and Betty received a grade of 92. What grade did Carol receive if the average of the three grades was 91? *Answer:*

D. *Draw a circle around the correct letter.*

28. Seventy-five percent of the students in a biology lab are men. If twenty men leave on a field trip, two-thirds of the remaining students are men. How many women are there in the lab?

    (A) 10        (B) 15        (C) 20        (D) 22        (E) 25

29. At a party 60% of the people are women. If 4 women leave, then 50% of the remaining people are women. How many women were there originally at the party?

    (A) 8         (B) 10        (C) 12        (D) 16        (E) 20

30. A basketball team wins 60% of its games. It then loses 8 games in a row. Altogether, it has won 50% of its games. How many games has the team won?

(A) 20          (B) 24          (C) 30          (D) 36          (E) 60

31. A baseball team has won 40% of its games. It then wins 10 games in a row. At that point, it has won 70% of its games. Altogether, how many games has the team played?

(A) 12          (B) 15          (C) 20          (D) 30          (E) 35

# 35.  Square Roots and Right Triangles

The Pythagorean Theorem for right triangles, which will be our final topic, requires a working ability with square roots.

## A. SQUARE ROOTS

You know that

$$2^2 = 2 \times 2 = 4$$

Suppose you are asked,

What *positive* number times itself is 4?

In other words, you are asked to find a *positive* number such that

(the number) $\times$ (the number) $= 4$

Clearly, 2 satisfies this condition. 2 is called the "positive square root of 4." Note that the *negative* number −2 also works because

$$(-2) \times (-2) = 4$$

−2 is called the "negative square root of 4."

DEFINITION

> SQUARE ROOT. Let $a$ be positive. Then $b$ is called the **positive square root of $a$** if $b$ is positive and if
>
> $$b^2 = a$$
>
> In this case, $-b$ is called the **negative square root of $a$**. Also the **square root of 0** is 0.

Write

$$b = \sqrt{a} \qquad Read: b \text{ equals the positive square root of } a.$$

if $b$ is the *positive* square root of $a$. Thus

$$\sqrt{4} = 2 \qquad \text{The } positive \text{ square root of 4 is 2.}$$

**Example 1** ▶   Find:

(a) $\sqrt{9}$      (b) $\sqrt{25}$      (c) $\sqrt{144}$

*Solution.*

(a) $\sqrt{9} = 3$  because 3 is positive and $3^2 = 9$.

(b) $\sqrt{25} = 5$  because 5 is positive and $5^2 = 25$.

(c) $\sqrt{144} = 12$  because 12 is positive and $12^2 = 144$.      ◀

To indicate that $-2$ is the *negative square root* of 4, write

$$-\sqrt{4} = -2$$

Similarly,

$$-\sqrt{9} = -3$$

Next, observe that

$$1^2 = 1 \qquad \text{and} \qquad 2^2 = 4$$

Is it possible to find a positive number $a$ whose *square* equals 2? In other words, can we find a positive number $a$ such that

$$a^2 = 2$$

and therefore

$$a = \sqrt{2}$$

Observe that

$$
\begin{array}{r}
1.41 \\
\times\ 1.41 \\
\hline
1\ 41 \\
56\ 4 \\
1\ 41 \\
\hline
1.98\ 81
\end{array}
\qquad
\begin{array}{r}
1.42 \\
\times\ 1.42 \\
\hline
2\ 84 \\
56\ 8 \\
1\ 42 \\
\hline
2.01\ 64
\end{array}
$$

Thus

$$(1.41)^2 = 1.9881 \text{ and } (1.42)^2 = 2.0164$$

It follows that

$$(1.41)^2 < 2 \quad \text{and} \quad 2 < (1.42)^2$$

and therefore that

$$1.41 < \sqrt{2} \quad \text{and} \quad \sqrt{2} < 1.42$$

Is there such a number $\sqrt{2}$? Consider the three squares in the figure below.

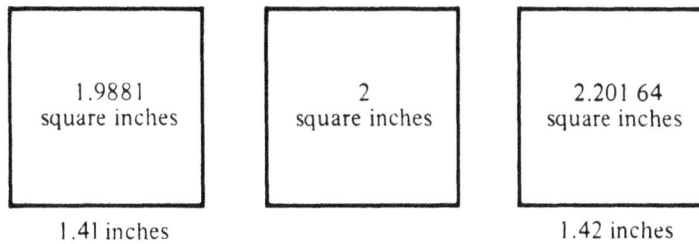

| 1.9881 square inches | 2 square inches | 2.201 64 square inches |
| 1.41 inches | | 1.42 inches |

With very precise instruments, we might be able to draw the first and third squares whose side lengths are 1.41 inches and 1.42 inches, respectively. Whether or not we can actually draw the middle square, doesn't it seem reasonable that there should be a square whose area is 2 square inches? The length of a side is then $\sqrt{2}$ inches.

It can be shown that $\sqrt{2}$ cannot be written as an "ordinary" decimal, no matter how many decimal digits we take. But for most purposes, it can be approximated closely enough by using

$$\sqrt{2} \approx 1.414 \qquad \textit{Read}: \sqrt{2} \text{ is approximately equal to 1.414.}$$

Similarly,

$$\sqrt{3} \approx 1.732$$

*Try the exercises for Topic A on page 258.*

## B. PRODUCTS AND QUOTIENTS OF ROOTS

Observe that

$$\sqrt{9 \times 4} = \sqrt{36} = 6$$

Furthermore,

$$\sqrt{9} \times \sqrt{4} = 3 \times 2 = 6$$

It follows that

$$\sqrt{9 \times 4} = \sqrt{9} \times \sqrt{4}$$

This suggests that (with suitable restrictions) the *square root of a product* equals the product of the square roots. More precisely, let a and b be positive. Then

$$\sqrt{ab} = \sqrt{a}\,\sqrt{b}$$

**Example 2** ▶    $\sqrt{2500} =$

(A) 5          (B) 50          (C) 500          (D) 62,500          (E) 6,250,000

*Solution.*  Let $a = 25$ and $b = 100$. Use the above rule.

$$\sqrt{2500} = \sqrt{25 \times 100}$$
$$= \sqrt{25} \times \sqrt{100}$$
$$= 5 \times 10$$
$$= 50$$

The correct choice is (B).                                                    ◀

   Sometimes we use the preceding rule in the reverse order. Thus, for positive numbers a and b,

$$\sqrt{a}\,\sqrt{b} = \sqrt{ab}$$

*The product of the roots is the root of the product.*

**Example 3** ▶    $\sqrt{18} \times \sqrt{2} =$

*Solution.*  Let $a = 18$ and $b = 2$. Use

$$\sqrt{a}\,\sqrt{b} = \sqrt{ab}$$
$$\sqrt{18} \times \sqrt{2} = \sqrt{\underset{\displaystyle \llcorner 36 \lrcorner}{18 \times 2}}$$
$$= \sqrt{36}$$
$$= 6$$                                                                          ◀

   Next, observe that

$$\sqrt{\frac{100}{25}} = \sqrt{4} = 2$$

Also,

$$\frac{\sqrt{100}}{\sqrt{25}} = \frac{10}{5} = 2$$

It follows that

$$\sqrt{\frac{100}{25}} = \frac{\sqrt{100}}{\sqrt{25}}$$

*The square root of a quotient equals the quotient of the square roots.* More precisely, if $a$ and $b$ are both positive, then

$$\sqrt{\frac{a}{b}} = \frac{\sqrt{a}}{\sqrt{b}}$$

**Example 4** ▶ $\sqrt{\frac{4}{9}} =$

(A) $\frac{16}{81}$     (B) $\frac{4}{81}$     (C) $\frac{4}{9}$     (D) $\frac{2}{9}$     (E) $\frac{2}{3}$

**Solution.** Let $a = 4$ and $b = 9$ in the preceding rule.

$$\sqrt{\frac{4}{9}} = \frac{\sqrt{4}}{\sqrt{9}} = \frac{2}{3}$$

The correct choice is (E). ◀

*Try the exercises for Topic B on page 258.*

## C. SQUARE ROOTS AND OTHER OPERATIONS

**Example 5** ▶ Find $\sqrt{9 + 16}$.

(A) 3     (B) 4     (C) 5     (D) 7     (E) 25

**Solution.** You are asked to find the (positive) square root of the sum, $9 + 16$. Thus first add 9 and 16; then find the square root of the sum.

$$\sqrt{9 + 16} = \sqrt{25} = 5$$
$$\underset{25}{\underbrace{\phantom{9 + 16}}}$$

The correct choice is (C). ◀

Observe that

$$\sqrt{9} = 3 \quad \text{and} \quad \sqrt{16} = 4$$

so that

$$\sqrt{9} + \sqrt{16} = 3 + 4 = 7$$

But, in Example 5 we saw that

$$\sqrt{9 + 16} = \sqrt{25} = 5$$

Therefore,

$$\sqrt{9 + 16} \neq \sqrt{9} + \sqrt{16}$$

(Read: $\sqrt{9 + 16}$ does *not* equal $\sqrt{9} + \sqrt{16}$.) In general, for positive numbers $a$ and $b$,

$$\sqrt{a+b} \neq \sqrt{a} + \sqrt{b}$$

and if $a > b$,

$$\sqrt{a-b} \neq \sqrt{a} - \sqrt{b}$$

To sum up, for positive numbers $a$ and $b$,

$$\sqrt{ab} = \sqrt{a}\,\sqrt{b} \qquad \text{and} \qquad \sqrt{\frac{a}{b}} = \frac{\sqrt{a}}{\sqrt{b}}$$

However,

$$\sqrt{a+b} \neq \sqrt{a} + \sqrt{b} \qquad \text{and} \qquad \sqrt{a-b} \neq \sqrt{a} - \sqrt{b}$$

Example 6 will illustrate a process that has an important application to geometry [*Topic* (D)].

**Example 6** ▶  Find    $\sqrt{10^2 - 6^2}$.

(A) 2    (B) 4    (C) 8    (D) 64    (E) 4096

*Solution.*  First square; then subtract; then find the square root of the difference.

$$\sqrt{10^2 - 6^2} = \sqrt{100 - 36}$$
$$= \sqrt{64}$$
$$= 8$$

The correct choice is (C).    ◀

*Try the exercises for Topic C on page 259.*

## D.  PYTHAGOREAN THEOREM

A **right triangle** is a triangle in which one of the angles is 90°. This angle is called a **right angle**. The side opposite the right angle is called the **hypotenuse**. The ancient Greeks discovered that the sides of lengths $a$, $b$, and $c$ of a right triangle are related by the formula

$$c^2 = a^2 + b^2$$

Here $c$ is the length of the hypotenuse. This relationship is known as the **Pythagorean Theorem**. (*See the figure below.*) Obtain the square root of each side of the above equation. Because $c$ is positive,

$$c = \sqrt{a^2 + b^2}$$

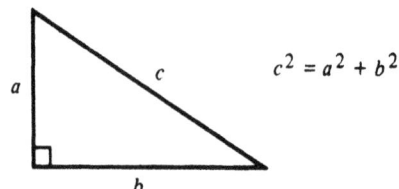
$$c^2 = a^2 + b^2$$

**Example 7** ▶  In the right triangle shown here, find $c$.

(A) 5        (B) 7        (C) 9

(D) 25      (E) 49

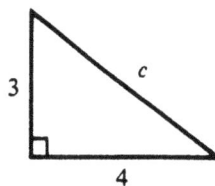

**Solution.**

$$c = \sqrt{a^2 + b^2}$$

Substitute 3 for $a$ and 4 for $b$.

$$c = \sqrt{3^2 + 4^2}$$
$$= \sqrt{9 + 16}$$
$$= \sqrt{25}$$
$$= 5$$

The correct choice is (A).                    ◀

**Example 8** ▶  In the right triangle shown here, find $b$.

**Solution.**  Here the right triangle is "turned around." We are given that $a = 5$ and that $c$, the length of the hypotenuse, is equal to 13.

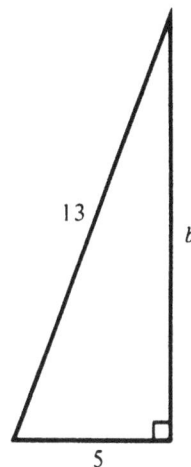

$$a^2 + b^2 = c^2 \qquad \text{Subtract } a^2 \text{ from each side.}$$
$$b^2 = c^2 - a^2 \qquad \text{Find the (positive) square root of each side.}$$

$$b = \sqrt{c^2 - a^2}$$

Substitute 13 for $c$ and 5 for $a$.

$$b = \sqrt{13^2 - 5^2}$$
$$= \sqrt{169 - 25}$$
$$= \sqrt{144}$$
$$= 12$$                                ◀

**Example 9** ▶  In the right triangle shown here, find $c$.

(A) 2        (B) 4        (C) 8

(D) 6        (E) $\sqrt{34}$

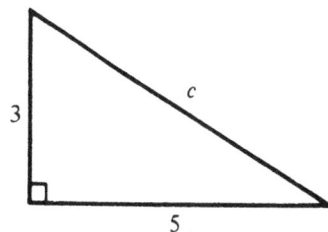

**Solution.**

$$c = \sqrt{a^2 + b^2}$$

Substitute 3 for $a$ and 5 for $b$.

$$c = \sqrt{3^2 + 5^2}$$

$$= \sqrt{9 + 25}$$
$$= \sqrt{34}$$

$\sqrt{34}$ is *less than* 6 because $\sqrt{36} = 6$. Thus the correct choice is (E).   ◀

*Try the exercises for Topic D on page 260.*

## EXERCISES

**A.** *Find each square root.*

1. $\sqrt{16}$     *Answer:*        2. $\sqrt{36}$     *Answer:*

3. $\sqrt{64}$     *Answer:*        4. $\sqrt{100}$     *Answer:*

5. $\sqrt{49}$     *Answer:*        6. $\sqrt{81}$     *Answer:*

7. $\sqrt{121}$     *Answer:*        8. $\sqrt{169}$     *Answer:*

9. $\sqrt{0}$     *Answer:*        10. $\sqrt{1}$     *Answer:*

**B.** *When several choices are given, draw a circle around the correct letter.*

11. $\sqrt{400} =$

    (A) 15     (B) 20     (C) 25     (D) 100     (E) 200

12. $\sqrt{900} =$

    (A) 3     (B) 30     (C) 300     (D) 450     (E) 90

13. $\sqrt{8100} =$

    (A) 9     (B) 81     (C) 90     (D) 900     (E) 4050

14. $\sqrt{10,000} =$

    (A) 10     (B) 100     (C) 1000     (D) 5000     (E) 100,000

15. $\sqrt{640,000} =$

16. $\sqrt{1,000,000} =$

17. $\sqrt{3} \times \sqrt{12}$

    (A) 15     (B) $\sqrt{15}$     (C) 36     (D) 6     (E) 216

18. $\sqrt{8} \times \sqrt{2} =$

    (A) 16     (B) 256     (C) 10     (D) $\sqrt{10}$     (E) 4

19. $\sqrt{5} \times \sqrt{5} =$

20. $\sqrt{20} \times \sqrt{5} =$

21. $\sqrt{32} \times \sqrt{2} =$

22. $\sqrt{48} \times \sqrt{3} =$

23. $\sqrt{\dfrac{1}{16}} =$

    (A) $\dfrac{1}{2}$       (B) $\dfrac{1}{4}$       (C) $\dfrac{1}{8}$       (D) 4       (E) $\dfrac{1}{256}$

24. $\sqrt{\dfrac{9}{25}} =$

    (A) $\dfrac{3}{25}$       (B) $\dfrac{9}{5}$       (C) $\dfrac{3}{5}$       (D) $\dfrac{81}{5}$       (E) $\dfrac{81}{25}$

25. $\sqrt{\dfrac{4}{81}} =$

26. $\sqrt{\dfrac{49}{100}} =$

27. $\sqrt{\dfrac{25}{64}} =$

28. $\sqrt{\dfrac{49}{4}} =$

29. $\sqrt{\dfrac{100}{81}} =$

30. $\sqrt{\dfrac{121}{144}} =$

**C.** *When several choices are given, draw a circle around the correct letter.*

31. $\sqrt{4 + 5} =$

    (A) $\sqrt{6}$       (B) 6       (C) 9       (D) 3       (E) $\sqrt{45}$

32. $\sqrt{7 + 18} =$

    (A) 25       (B) 5       (C) 125       (D) $\sqrt{125}$       (E) $\sqrt{7} + \sqrt{18}$

33. $\sqrt{16 - 7} =$

    (A) 9       (B) 3       (C) $\sqrt{16} - \sqrt{7}$       (D) $\sqrt{\dfrac{16}{7}}$       (E) 81

**34.** $\sqrt{100 - 64} =$

     (A) 2        (B) 4        (C) 6        (D) 8        (E) 36

**35.** $\sqrt{4 + 9} =$

**36.** $\sqrt{49 - 36} =$

**37.** $\sqrt{3^2 + 4^2} =$

     (A) 25        (B) 125        (C) 5        (D) 7        (E) 49

**38.** $\sqrt{8^2 + 6^2} =$

     (A) 10        (B) 100        (C) 14        (D) 48        (E) $\sqrt{48}$

**39.** $\sqrt{10^2 - 8^2} =$

     (A) 2        (B) 4        (C) 16        (D) 6        (E) 36

**40.** $\sqrt{13^2 - 12^2} =$

     (A) 11        (B) 121        (C) 5        (D) 25        (E) 125

**41.** $\sqrt{4^2 + 5^2} =$

**42.** $\sqrt{4^2 - 3^2} =$

**D.** *Find the length of the indicated side of the right triangle. Draw a circle around the correct letter.*

**43.**

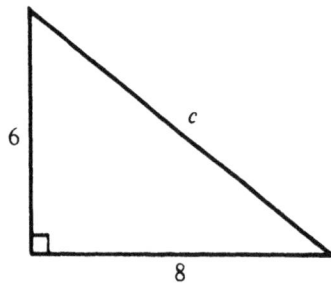

     (A) 10     (B) 12     (C) 14

     (D) 2      (E) 196

**44.**

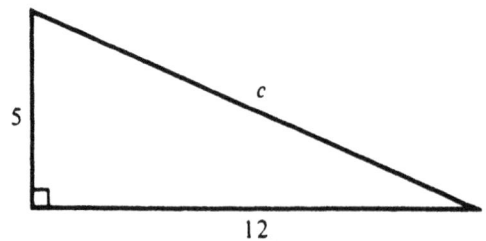

     (A) 13     (B) 15     (C) 17

     (D) 20     (E) 25

**45.**

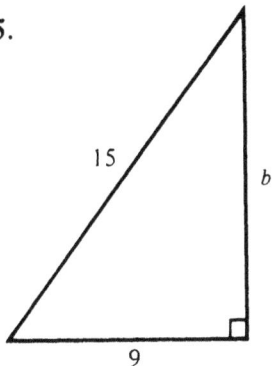

(A) 6      (B) 9      (C) 10

(D) 12      (E) 24

**46.**

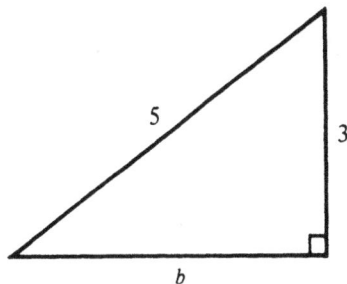

(A) 2      (B) 3      (C) 4

(D) $\sqrt{3}$      (E) 8

**47.**

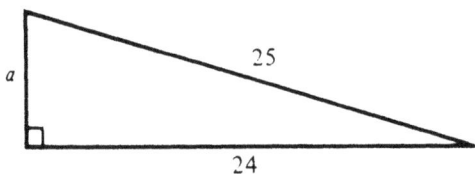

(A) 1      (B) 2      (C) 5

(D) 7      (E) 12

**48.**

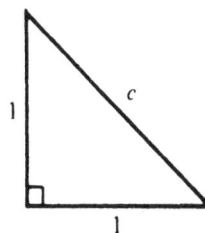

(A) 1      (B) 2      (C) 3

(D) $\sqrt{2}$      (E) $\sqrt{3}$

**49.**

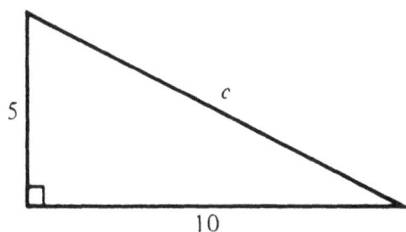

(A) 5      (B) 10      (C) 12

(D) 15      (E) $\sqrt{125}$

**50.**

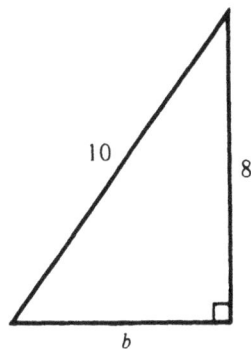

(A) 2      (B) 4      (C) 6

(D) $\sqrt{6}$      (E) 36

**51.**

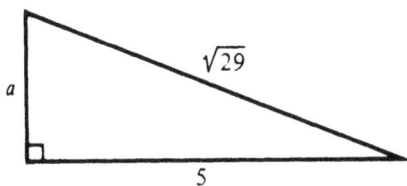

(A) 2      (B) 4      (C) 8

(D) 24      (E) $\sqrt{24}$

**52.**

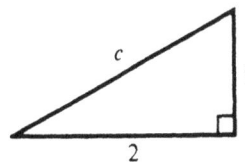

(A) 3      (B) 4      (C) 5

(D) $\sqrt{5}$      (E) 25

# Exam A on Algebra

*When several choices are given, draw a circle around the correct letter.*

1. Find the sum of $3x + 6y$ and $4x - 7y$.    *Answer:*

2. What is the product of $10a^2bc^3$ and $-2a^2c$?    *Answer:*

3. Simplify $5x^2 - 4x(x + 2)$.    *Answer:*

4. Add $6a^3 + 3a$ and $9 - 3a$.    *Answer:*

5. $(3x^2y^3)^3 =$

   (A) $9x^6y^9$            (B) $9x^5y^6$            (C) $27x^5y^6$

   (D) $27x^6y^9$           (E) $27x^5y^9$

6. Simplify $\dfrac{4a^2 + 20a}{2a}$.

   (A) $2a + 20$           (B) $2a + 10$           (C) $22a$

   (D) $2a + 10a$          (E) $2a + 20a$

7. Factor $9a^2 - 18a$.    *Answer:*

8. Find $x$ if $4x - 3 = 3x + 2$.    *Answer:*

9. Find the value, in cents, of $q$ quarters and $d$ dimes.

   (A) $q + d$             (B) $25(q + d)$         (C) $25(q + 10d)$

   (D) $25q + 10d$         (E) $\dfrac{q}{25} + \dfrac{d}{10}$

10. If $x = 3$ and $y = -4$, find the value of $x^2 - 2xy$.    *Answer:*

11.  If $\frac{x}{5} = \frac{7}{2}$, find $x$.     *Answer:*

12.  If $c = -3ab^2$, find $c$ when $a = 4$ and $b = -1$.     *Answer:*

13.  If $a = 5b - 3$, then $b =$

(A)  $5a + 3$         (B)  $5a + 15$      (C)  $a + \frac{3}{5}$        (D)  $\frac{a + 3}{5}$       (E)  $\frac{a - 3}{5}$

14.  Which of these points lies on the graph of $4x - 3y = 9$.

(A)  $(2, 1)$              (B)  $(5, -3)$          (C)  $(3, 1)$            (D)  $(3, -1)$           (E)  $(1, -3)$

15.  If $(0, b)$ lies on the graph of $y = 10x - 7$, then $b =$

(A)  7                    (B)  10                  (C)  $-7$

(D)  $(0, 7)$            (E)  $(0, -7)$

16.  An equation for this line is

(A)  $x + y = 2$        (B)  $x + y = -2$      (C)  $x - y - 2 = 0$

(D)  $2x - y = 2$      (E)  $2y - x = 2$

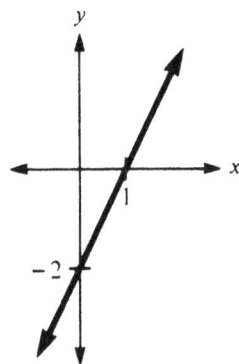

17.  A bookstore finds that 2 out of every 9 books sold are returned. If 360 books are sold one week, how many of these are expected to be returned?

(A)  40                  (B)  80                  (C)  180                (D)  162                (E)  1620

18.  In the right triangle shown here, find $x$.

(A)  3              (B)  $\sqrt{3}$             (C)  9

(D)  81             (E)  $\sqrt{369}$

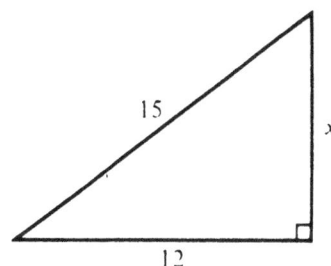

19.  If $3x - y = 10$ and $4x + y = 11$, then

(A)  $x = 4, y = 2$              (B)  $x = 5, y = 5$                (C)  $x = 0, y = -10$

(D)  $x = 3, y = 1$              (E)  $x = 3, y = -1$

20.  Tickets for a concert sell for $2.50 each. Expenses for the concert amount to $3000. If a profit of $2000 is made, how many tickets are sold?

(A)  800                (B)  1200               (C)  2000               (D)  3000               (E)  5000

# Exam B on Algebra

*When several choices are given, draw a circle around the correct letter.*

1. The sum of $10x - 2y$ and $-2x + y$ is

   (A) $8x + y$     (B) $8x - 2y$     (C) $8x - y$     (D) $12x - y$     (E) $12x + 3y$

2. Multiply $-2xy^2$ by $4x^3y^2$.  *Answer:*

3. Simplify $5a^2b - 4ab(a - b)$.

   (A) $a^2b - b$                  (B) $a^2b + 4ab^2$                  (C) $9a^2b - 4ab^2$

   (D) $9a^2b - 4ab$               (E) $a^2b + 4a^2b$

4. Subtract $4x - 6$ from $6x^2 - 12$.    *Answer:*

5. Simplify $\dfrac{24a^2 - 4a}{4a}$.   *Answer:*

6. $(4p^2q)^3 =$

   (A) $16p^5q^3$     (B) $16p^6q^3$     (C) $64p^5q^3$     (D) $64p^6q^3$     (E) $64p^2q^3$

7. The total value, in cents, of $n$ nickels and $q$ quarters is

   (A) $n + q$                  (B) $5n + q$                  (C) $5(n + q)$

   (D) $5(n + 25q)$             (E) $5n + 25q$

8. Factor $6a^2b - 27ab$ completely.   *Answer:*

9. If $4x - 3 = 2x + 7$, then $x =$

   (A) 2          (B) 4          (C) 5          (D) 6          (E) $-3$

10. If $\dfrac{x}{4} - 1 = \dfrac{x}{6}$, then $x =$

    (A) 4          (B) 8          (C) 12          (D) $-12$          (E) 24

11. If $\dfrac{x+2}{5} = \dfrac{x-4}{2}$, then $x =$

(A) 2          (B) 4          (C) 8          (D) 10          (E) 20

12. Which of the following points lies on the graph of $2x - 3y = 10$?

(A) $(1, 1)$       (B) $(2, 2)$       (C) $(2, -2)$       (D) $(3, -2)$       (E) $(2, -4)$

13. What is the value of $3a^2 + 4ab$ when $a = 2$ and $b = 4$?   *Answer:*

14. An equation for the line shown here is

(A) $x + y = 0$               (B) $y = 2x$

(C) $x - 2y = 0$              (D) $y = x + 2$

(E) $y = x - 2$

15. If $5x - 9 = y$, then $x =$

(A) $y + 9$       (B) $\dfrac{y+9}{5}$       (C) $\dfrac{y-9}{5}$       (D) $5y + 45$       (E) $\dfrac{y}{5} + 9$

16. A typist earns \$70 in 8 hours. At this rate of pay, how much will she earn in 12 hours?

(A) \$74          (B) \$100          (C) \$105          (D) \$110          (E) \$46.67

17. A taxi company owns 50 cabs, 40% of which are air-conditioned. How many additional air-conditioned cabs must be purchased in order that 50% of the cabs should then be air-conditioned? (All of the old cabs are retained.)

(A) 10          (B) 20          (C) 25          (D) 40          (E) 50

18. The graphs of $4x + 2y = 8$ and $5x - 2y = 1$ meet at a point where

(A) $x = 2, y = 0$               (B) $x = 0, y = 4$               (C) $x = 2, y = 2$

(D) $x = 2, y = 1$               (E) $x = 1, y = 2$

19. Rose has grades of 94 and 89 on her first two math exams. What grade must she get on her third exam in order to have a 90 average on her three exams?

(A) 87          (B) 88          (C) 90          (D) 91          (E) 93

20. Among 105 workers in a factory, the number of women is 15 more than twice the number of men. Find the number of women who work in the factory.

(A) 30          (B) 65          (C) 70          (D) 75          (E) 100

# Sample Final Exams

You are now given three sample exams for practice. Complete solutions to these sample exam questions are given, beginning on page 292.

## Sample Final Exam A

*Choose the answer you think is correct. Then, in the answer column at the right, fill in the box underneath the corresponding letter.*

1. Four hundred twenty thousand five hundred is written

   (A) 420,005　　　(B) 425,000　　　(C) 425,500

   (D) 420,500　　　(E) 420,000.05

2. $328 - 79 =$

   (A) 249　(B) 259　(C) 407　(D) 149　(E) 159

3. $6512 \div 16 =$

   (A) 470　(B) 407　(C) 47　(D) 40　(E) 40.7

4. The average of 78, 81, 82, and 91 is

   (A) 80　(B) 82.5　(C) 83　(D) 84　(E) 85

5. A club sells 952 raffles at $3 each. The printing of the raffle books costs $97. If there are $52 in other expenses, the profit is

   (A) $2856　　(B) $2707　　(C) $2717　　(D) $3005　　(E) $2759

```
    A B C D E
1   ▯ ▯ ▯ ▯ ▯
    A B C D E
2   ▯ ▯ ▯ ▯ ▯
    A B C D E
3   ▯ ▯ ▯ ▯ ▯
    A B C D E
4   ▯ ▯ ▯ ▯ ▯
    A B C D E
5   ▯ ▯ ▯ ▯ ▯
```

266

6. Which of these fractions is the smallest?

(A) $\frac{2}{3}$    (B) $\frac{3}{5}$    (C) $\frac{3}{4}$    (D) $\frac{5}{8}$    (E) $\frac{5}{6}$

7.    27 pounds 12 ounces
    −18 pounds 14 ounces

(A) 9 pounds 2 ounces        (B) 8 pounds 2 ounces

(C) 8 pounds 8 ounces        (D) 8 pounds 14 ounces

(E) 9 pounds 14 ounces

8. $\frac{2}{9} + \frac{3}{4} =$

(A) $\frac{5}{13}$    (B) $\frac{5}{36}$    (C) $\frac{35}{36}$    (D) $\frac{36}{35}$    (E) $\frac{11}{36}$

9. $4\frac{1}{2} - 2\frac{1}{3} =$

(A) $2\frac{1}{3}$    (B) $2\frac{1}{6}$    (C) $1\frac{1}{6}$    (D) $2\frac{1}{5}$    (E) $1\frac{1}{5}$

10. $8\frac{1}{4} \div 2\frac{3}{4} =$

(A) 3    (B) $3\frac{1}{2}$    (C) 4    (D) $\frac{4}{33}$    (E) $\frac{363}{16}$

11. 15.093 + 2.979 + 8 =

(A) 25.072        (B) 26.072        (C) 26.062

(D) 25.962        (E) 26.909

12. Which of these numbers is the smallest?

(A) .0404    (B) .0440    (C) .0309    (D) .0490    (E) .0390

13. 87.8 − 7.96 =

(A) 79.94    (B) 79.84    (C) 8.2    (D) .82    (E) 8.22

14. Change $\frac{12}{13}$ to a decimal rounded to the nearest hundredth.

(A) .92    (B) .923    (C) .93    (D) .9    (E) .90

15. A store sells 16 copies of a textbook priced at $7.95. What are the total sales for this book?

(A) $102    (B) $128    (C) $127.20    (D) $12.70    (E) $128.20

16. What is 60% of 75?

(A) 15    (B) 25    (C) 40    (D) 45    (E) 50

17. If 60 is 80% of a certain number, find that number.

(A) 48    (B) 480    (C) 72    (D) 75    (E) 90

6   A B C D E
7   A B C D E
8   A B C D E
9   A B C D E
10  A B C D E
11  A B C D E
12  A B C D E
13  A B C D E
14  A B C D E
15  A B C D E
16  A B C D E
17  A B C D E

18. If a $45 dress is on sale reduced by 10%, its sale price is

(A) $4.50      (B) $49.50      (C) $40      (D) $40.50      (E) $41.50

19. Find the area of a rectangular table top that is 40 inches long and 22 inches wide.

(A) 800 in.²      (B) 880 in.²      (C) 62 in.      (D) 124 in.      (E) 808 in.²

20. Use the graph at the right to find the total number of copies of Health Digest sold from September through December, 1982.

1982 Sales
(Thousands)

(A) 11,000      (B) 12,000

(C) 12,500      (D) 13,000

(E) 14,000

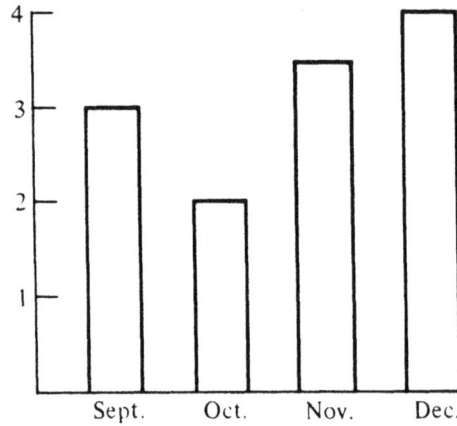

| | A | B | C | D | E |
|---|---|---|---|---|---|
| 18 | 0 | 0 | 0 | 0 | 0 |
| 19 | 0 | 0 | 0 | 0 | 0 |
| 20 | 0 | 0 | 0 | 0 | 0 |
| 21 | 0 | 0 | 0 | 0 | 0 |
| 22 | 0 | 0 | 0 | 0 | 0 |
| 23 | 0 | 0 | 0 | 0 | 0 |
| 24 | 0 | 0 | 0 | 0 | 0 |
| 25 | 0 | 0 | 0 | 0 | 0 |

21. Find the sum of $4x + 9y$ and $6x - 9y$.

(A) $10x$      (B) $18y$      (C) $10x - 18y$

(D) $10x + 9y$      (E) $2x - 18y$

22. What is the product of $6a^2bc^3$ and $-2ab^4c^2$?

(A) $4a^3b^5c^5$      (B) $-12a^3b^5c^5$      (C) $-12a^2b^4c^6$

(D) $-12a^3b^5c^6$      (E) $12a^3b^5c^5$

23. Simplify $5a^2 - 4a(a - 3)$.

(A) $a^2 - 3$      (B) $a^2 - 4a$      (C) $a^2 - 12a$

(D) $a^2 + 12a$      (E) $13a^2$

24. Add $3x^2 - 2x + 4$ and $x^2 - 5$.

(A) $4x^2 - 2x - 1$      (B) $4x^2 - 2x + 1$      (C) $4x^2 + 2x - 1$

(D) $4x^2 - 1$      (E) $2x^2 - 2x + 9$

25. $(5x^3y^4)^2 =$

(A) $25x^6y^8$      (B) $25x^3y^4$      (C) $25x^5y^6$

(D) $5x^6y^8$      (E) $10x^6y^8$

26. Simplify $\dfrac{12a^2 + 16a}{-4a}$.

    (A) $3a + 16$      (B) $3a + 4$      (C) $3a - 4$

    (D) $-3a - 4$      (E) $-3a - 16$

27. Factor $9x^3 - 15x^2$.

    (A) $x(9x^2 - 15x)$      (B) $x^2(9x - 15)$      (C) $9x^2(x - 15)$

    (D) $3x(3x^2 - 5)$      (E) $3x^2(3x - 5)$

28. $2(-5)^2 + 4(-7) =$

    (A) $-78$      (B) $22$      (C) $72$      (D) $6$      (E) $-128$

29. Find $x$ if $4x + 3 = 5x - 1$.

    (A) $1$      (B) $2$      (C) $3$      (D) $4$      (E) $5$

30. Find the value, in dollars, of $t$ ten-dollar bills and $f$ five-dollar bills.

    (A) $f + t$      (B) $50\,ft$      (C) $5f + 10t$

    (D) $5(f + 10t)$      (E) $f + 10t$

31. If $x = 4$ and $y = -3$, find the value of $x^2 - 5xy$.

    (A) $76$      (B) $-54$      (C) $-44$      (D) $-51$      (E) $69$

32. If $\dfrac{x}{5} = \dfrac{-3}{4}$, then $x =$

    (A) $15$      (B) $-15$      (C) $-\dfrac{15}{4}$      (D) $-\dfrac{3}{20}$      (E) $-\dfrac{20}{3}$

33. Which of these points lies on the graph of $4x - y = 9$?

    (A) $(2, 1)$      (B) $(2, -1)$      (C) $(0, 9)$      (D) $(1, 5)$      (E) $(3, -3)$

34. An equation for the line at the right is

    (A) $x = -1$      (B) $y = 2$      (C) $y - 2x = 1$

    (D) $y - 2x = 2$      (E) $x - 2y = 2$

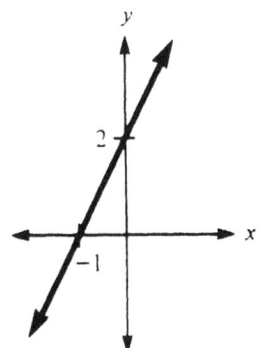

**Answer grid:**

|    | A | B | C | D | E |
|----|---|---|---|---|---|
| 26 | [] | [] | [] | [] | [] |
| 27 | [] | [] | [] | [] | [] |
| 28 | [] | [] | [] | [] | [] |
| 29 | [] | [] | [] | [] | [] |
| 30 | [] | [] | [] | [] | [] |
| 31 | [] | [] | [] | [] | [] |
| 32 | [] | [] | [] | [] | [] |
| 33 | [] | [] | [] | [] | [] |
| 34 | [] | [] | [] | [] | [] |

35.  If $c = 8ab^2$, find $c$ when $a = -1$ and $b = 2$.

(A) $-16$     (B) $16$     (C) $-32$     (D) $32$     (E) $256$

36.  If $4p - q = 5$, then $p =$

(A) $q + 5$          (B) $q - 5$          (C) $\dfrac{q + 5}{4}$

(D) $\dfrac{q}{4} + 5$          (E) $q + \dfrac{5}{4}$

37.  A batter gets 2 hits for every 7 times at bat. How many times must he come to bat in order to have 150 hits?

(A) 350     (B) 420     (C) 500     (D) 525     (E) 550

38.  In the right triangle shown here, find $x$.

(A) 5          (B) 17          (C) 289

(D) $\sqrt{17}$          (E) $\sqrt{15}$

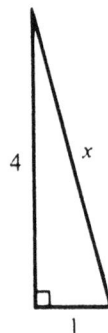

39.  If $3x + 2y = 7$ and $3x - y = 1$, then

(A) $x = 1$ and $y = 1$     (B) $x = 1$ and $y = 2$     (C) $x = 2$ and $y = 1$

(D) $x = 3$ and $y = -1$     (E) $x = 2$ and $y = 5$

40.  It costs a toy manufacturer $5 to produce a hockey game. In addition, there is a general overhead cost of $15,000. He decides to produce 6000 games. If he receives $15 per game from a wholesaler, how many games must he sell to break even.

(A) 1000          (B) 3000          (C) 4000

(D) 4500          (E) 5000

## Sample Final Exam B

*Choose the answer you think is correct. Then, in the answer column at the right, fill in the box underneath the corresponding letter.*

1. Sixty thousand four hundred five is written

    (A) 60,045　　　　(B) 64,105　　　　(C) 60,405

    (D) 64,405　　　　(E) 60,045

2. $6061 \div 19 =$

    (A) 309　　(B) 319　　(C) 39　　(D) 320　　(E) 31.9

3. 　10 hours   6 minutes
   − 4 hours 51 minutes

    (A) 6 hours 55 minutes　　　　(B) 5 hours 55 minutes

    (C) 6 hours 15 minutes　　　　(D) 5 hours 15 minutes

    (E) 5 hours 45 minutes

4. $\frac{1}{8} + \frac{3}{5} =$

    (A) $\frac{4}{13}$　　(B) $\frac{27}{40}$　　(C) $\frac{4}{40}$　　(D) $\frac{1}{10}$　　(E) $\frac{29}{40}$

5. If pencils sell for 15¢ each, how many can you buy for $6?

    (A) 90　　(B) 4　　(C) 40　　(D) 400　　(E) 900

6. $2\frac{3}{4} + 1\frac{1}{4} =$

    (A) $3\frac{3}{4}$　　(B) $3\frac{4}{5}$　　(C) 3　　(D) 4　　(E) 5

7. $3.09 - 1.98 =$

    (A) 2.07　　(B) 2.11　　(C) 1.11　　(D) 1.91　　(E) .11

8. It costs a food vendor 65¢ to make up each sandwich. He makes up 80 sandwiches and sells 70 for $1.50 each. What is his profit?

   (A) $59.50          (B) $68          (C) $52

   (D) $105            (E) $53

9. What is 9% of 8?

   (A) 72      (B) 7.2      (C) .72      (D) .072      (E) $\frac{8}{9}$

10. A sweater that normally sells for $32 is reduced in price by 25%. Its new price is

    (A) $8      (B) $24      (C) $7      (D) $25      (E) $25.60

11. If 8 ounces of a cheese costs $2.80, how much will a 12-ounce slice of this cheese cost?

    (A) $4.20          (B) $3.90          (C) $4.50

    (D) $5.60          (E) $3.60

12. On four math exams Sheila's scores are 72, 85, 80, and 87. What is her exam average?

    (A) 78      (B) 79      (C) 80      (D) 81      (E) 82

13. $4\frac{1}{2} \div 2\frac{1}{4} =$

    (A) $2\frac{1}{4}$      (B) 2      (C) $\frac{1}{2}$      (D) $\frac{1}{4}$      (E) $1\frac{3}{4}$

14. Which of the following fractions is the smallest?

    (A) $\frac{1}{7}$      (B) $\frac{1}{10}$      (C) $\frac{2}{11}$      (D) $\frac{2}{15}$      (E) $\frac{3}{20}$

15. Which of the following numbers is the smallest?

    (A) 2.04          (B) 2.39          (C) 2.009

    (D) 2.102         (E) 2.010

8  A B C D E
9  A B C D E
10 A B C D E
11 A B C D E
12 A B C D E
13 A B C D E
14 A B C D E
15 A B C D E

16.  $5.08 + 3.038 + 12 =$

(A) 20.46          (B) 23.16          (C) 21.18

(D) 20.118         (E) 20.046

17.  A jacket sells for \$85 plus a 6% sales tax. What is the total price?

(A) \$90           (B) \$90.10        (C) \$91

(D) \$5.10         (E) \$51

18.  A laborer earns \$44 for 8 hours of work. At this rate how much will he earn for 28 hours of work?

(A) \$144          (B) \$154          (C) \$164

(D) \$140          (E) \$180

19.  Oaktag sells for \$1.50 per square yard. What is the cost of a piece that is 6 yards by 5 yards?

(A) \$3     (B) \$40     (C) \$45     (D) \$20     (E) \$60

20.  The graph at the right shows the number of automobiles sold by a dealer during the first four months of the year. The *total* number sold during this four-month period is closest to

(A) 30             (B) 40             (C) 45

(D) 48             (E) 55

Automobile Sales

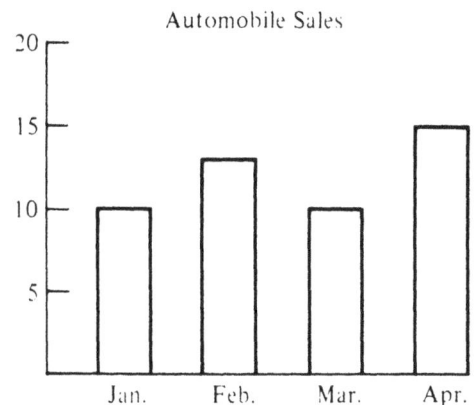

21.  The sum of $4a^2 + a - 3$ and $a + 2$ is

(A) $4a^2 + a + 5$        (B) $4a^2 + 2a - 5$        (C) $4a^2 + 2a - 1$

(D) $4a^2 + 2a + 1$       (E) $4a^2 - 1$

22.  Multiply $-5ab$ by $6a^2b^3$.

(A) $a^3b^4$             (B) $-30a^2b^3$            (C) $30a^3b^4$

(D) $-30a^3b^4$          (E) $-30a^3b^3$

23.  Simplify $2x^2y - 2x^2(y + 3)$.

(A) $4x^2y + 6x$          (B) $6x^2$                (C) $-6x^2$

(D) $4x^2y + 3$           (E) 3

16  A B C D E
17  A B C D E
18  A B C D E
19  A B C D E
20  A B C D E
21  A B C D E
22  A B C D E
23  A B C D E

24. Subtract $4a - 5$ from $5a + 4$.

(A) $9a - 1$         (B) $a - 1$         (C) $a - 9$

(D) $a + 9$          (E) $9a + 9$

25. Simplify $\dfrac{4a^2 - 12a}{2a}$.

(A) $2a$             (B) $-6$            (C) $2a + 6$

(D) $2a - 6$         (E) $-4a$

26. $(3x^4y^2)^3 =$

(A) $9x^{12}y^6$     (B) $9x^7y^5$       (C) $27x^7y^5$

(D) $27x^{12}y^5$    (E) $27x^{12}y^6$

27. If the sales tax rate is 4% and if the sales tax on a table is $6, what is the price of the table (before the sales tax)?

(A) $24            (B) $240           (C) $100

(D) $150           (E) $250

28. Suppose that gloves cost $9 a pair and scarves cost $7 apiece. The cost of $x$ pairs of gloves and $y$ scarves is

(A) $x + y$          (B) $9x + y$        (C) $9(x + y)$

(D) $9x + 7y$        (E) $9(x + 7y)$

29. Factor $25x^2y - 45xy$ completely.

(A) $5(5x^2y - 9xy)$   (B) $5x(5xy - 9y)$   (C) $25xy(x - 9y)$

(D) $5xy(5x - 9y)$     (E) $5xy(5x - 9)$

30. $(-2)(-5) - (-3)^2 =$

(A) 19       (B) 1       (C) $-19$       (D) 49       (E) 169

31. If $\dfrac{x + 3}{2} = \dfrac{3x - 3}{3}$, then $x =$

(A) 1       (B) 3       (C) 5       (D) $-5$       (E) 7

32. Which of the following points lies on the graph of $3x - 2y = 20$?

(A) $(4, 2)$         (B) $(6, 1)$        (C) $(10, 1)$

(D) $(8, 2)$         (E) $(8, -2)$

33. What is the value of $5ab - b^2$ when $a = 4$ and $b = -1$?

(A) 24       (B) $-21$       (C) $-19$       (D) 19       (E) 21

| | A | B | C | D | E |
|---|---|---|---|---|---|
| 24 | ▯ | ▯ | ▯ | ▯ | ▯ |
| 25 | ▯ | ▯ | ▯ | ▯• | ▯ |
| 26 | ▯ | ▯ | ▯ | ▯ | ▯ |
| 27 | ▯ | ▯ | ▯ | ▯ | ▯ |
| 28 | ▯ | ▯ | ▯ | ▯ | ▯ |
| 29 | ▯ | ▯ | ▯ | ▯ | ▯ |
| 30 | ▯ | ▯ | ▯ | ▯ | ▯ |
| 31 | ▯ | ▯ | ▯ | ▯ | ▯ |
| 32 | ▯ | ▯ | ▯ | ▯ | ▯ |
| 33 | ▯ | ▯ | ▯ | ▯ | ▯ |

34. An equation for the line at the right is

(A) $2x + 3y = 1$        (B) $2x = 3y$

(C) $3x = 2y$        (D) $2x + 3y = 6$

(E) $3x + 2y = 6$

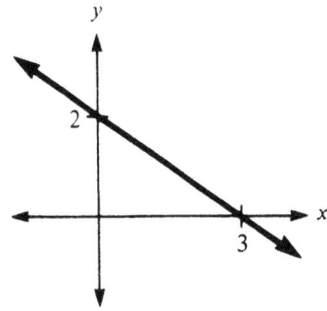

35. If $5x - 2 = 22$, then $x =$

(A) 4        (B) 5        (C) $\frac{24}{5}$        (D) 120        (E) 100

36. If $4a - b = c$, then $a =$

(A) $b + c$        (B) $4b + c$        (C) $4(b + c)$

(D) $\frac{b + c}{4}$        (E) $b + \frac{c}{4}$

37. The graphs of $2x + y = 5$ and $5x - y = 9$ meet at the point where

(A) $x = 2, y = 1$        (B) $x = 2, y = -1$        (C) $x = 1, y = -4$

(D) $x = 3, y = 6$        (E) $x = 0, y = -9$

38. The width and length of a 5 × 7 photograph are enlarged proportionally. If the width of the enlargement is $7\frac{1}{2}$ inches, what is the length?

(A) 9 inches        (B) 10 inches        (C) $10\frac{1}{2}$ inches

(D) 11 inches        (E) 12 inches

39. Simplify $\sqrt{10^2 - 6^2}$.

(A) 16        (B) 4        (C) 64        (D) 8        (E) $\sqrt{8}$

40. Expenses for a dance amount to $1260, and tickets to the dance sell for $8 each. If a profit of $1156 is made, how many tickets are sold?

(A) 200        (B) 144        (C) 300        (D) 302        (E) 312

| | A | B | C | D | E |
|---|---|---|---|---|---|
| 34 | ▯ | ▯ | ▯ | ▯ | ▯ |
| 35 | ▯ | ▯ | ▯ | ▯ | ▯ |
| 36 | ▯ | ▯ | ▯ | ▯ | ▯ |
| 37 | ▯ | ▯ | ▯ | ▯ | ▯ |
| 38 | ▯ | ▯ | ▯ | ▯ | ▯ |
| 39 | ▯ | ▯ | ▯ | ▯ | ▯ |
| 40 | ▯ | ▯ | ▯ | ▯ | ▯ |

# Sample Final Exam C

*Choose the answer you think is correct. Then, in the answer column at the right, fill in the box underneath the corresponding letter.*

| | A | B | C | D | E |
|---|---|---|---|---|---|
| 1 | ▯ | ▯ | ▯ | ▯ | ▯ |

1. Eighty thousand and four-tenths is written

(A) 80,004        (B) 80,410        (C) 80,000.4

(D) 80,000.04        (E) 80,004.1

2. $602 - 98 =$

(A) 604    (B) 504    (C) 598    (D) 584    (E) 700

3. The average of 10, 11, 12, and 13 is

(A) 11    (B) 11.5    (C) 12    (D) 12.5    (E) 46

4.    10 pounds 11 ounces
    − 6 pounds 12 ounces

(A) 4 pounds 1 ounce        (B) 3 pounds 1 ounce

(C) 3 pounds 9 ounces       (D) 3 pounds 15 ounces

(E) 4 pounds 15 ounces

5. $\frac{3}{8} - \frac{2}{5} =$

(A) $\frac{1}{3}$    (B) $\frac{1}{8}$    (C) $\frac{1}{40}$    (D) $-\frac{1}{40}$    (E) 0

6. $\frac{3}{5} \times \frac{10}{9} =$

(A) $\frac{13}{45}$    (B) $\frac{2}{3}$    (C) $\frac{13}{14}$    (D) $\frac{27}{50}$    (E) $\frac{30}{14}$

7. $\frac{3}{10} \div \frac{9}{20} =$

(A) $\frac{2}{3}$    (B) $\frac{3}{2}$    (C) $\frac{27}{200}$    (D) $\frac{2}{5}$    (E) 3

8. $3\frac{3}{4} + 1\frac{1}{4} =$

(A) 4    (B) 5    (C) $4\frac{1}{2}$    (D) $4\frac{3}{4}$    (E) $5\frac{1}{4}$

9. Which of these fractions is the smallest?

(A) $\frac{1}{2}$    (B) $\frac{4}{9}$    (C) $\frac{5}{11}$    (D) $\frac{5}{12}$    (E) $\frac{9}{20}$

10. Which of these numbers is the smallest?

(A) .0303        (B) .0299        (C) .0312

(D) .0289        (E) .0298

| | A | B | C | D | E |
|---|---|---|---|---|---|
| 2 | ▯ | ▯ | ▯ | ▯ | ▯ |
| 3 | ▯ | ▯ | ▯ | ▯ | ▯ |
| 4 | ▯ | ▯ | ▯ | ▯ | ▯ |
| 5 | ▯ | ▯ | ▯ | ▯ | ▯ |
| 6 | ▯ | ▯ | ▯ | ▯ | ▯ |
| 7 | ▯ | ▯ | ▯ | ▯ | ▯ |
| 8 | ▯ | ▯ | ▯ | ▯ | ▯ |
| 9 | ▯ | ▯ | ▯ | ▯ | ▯ |
| 10 | ▯ | ▯ | ▯ | ▯ | ▯ |

11.  $6.003 + 5.809 + 11 =$

    (A) 23.109       (B) 22.812       (C) 23.812

    (D) 22.839       (E) 22.849

12.  A gallon of gasoline costs \$1.25. What is the cost of 16 gallons of gasoline?

    (A) \$16       (B) \$18       (C) \$19

    (D) \$20       (E) \$20.50

13.  Find the cost of 5 pounds of potatoes at 22 cents per pound and 3 pounds of onions at 19 cents per pound.

    (A) \$1.10       (B) \$.57       (C) \$1.67

    (D) \$2.28       (E) \$3.28

14.  Change $\frac{4}{13}$ to a decimal rounded to the nearest hundredth.

    (A) .04    (B) .30    (C) .307    (D) .31    (E) .37

15.  If 48 is 75% of a certain number, find that number.

    (A) 36    (B) 3600    (C) 12    (D) 16    (E) 64

16.  An alloy contains 30% nickel. How much nickel is there in 120 tons of the alloy?

    (A) 30 tons       (B) 36 tons       (C) 360 tons

    (D) 40 tons       (E) 400 tons

17.  A radio, which was selling for \$60, was reduced by 25%. The new price is

    (A) \$40    (B) \$35    (C) \$45    (D) \$48    (E) \$50

18.  How much does it cost to carpet a room that is 16 feet by 12 feet, if carpeting costs \$4 per square foot?

    (A) \$192    (B) \$48    (C) \$64    (D) \$384    (E) \$768

19.  $(-2 - 1)^2 - 4(-3) =$

    (A) −8    (B) 16    (C) −3    (D) 21    (E) −21

Answer grid:

| | A | B | C | D | E |
|---|---|---|---|---|---|
| 11 | | | | | |
| 12 | | | | | |
| 13 | | | | | |
| 14 | | | | | |
| 15 | | | | | |
| 16 | | | | | |
| 17 | | | | | |
| 18 | | | | | |
| 19 | | | | | |

20. The graph at the right indicates the number of students enrolled at Southwestern Tech from 1977 to 1980. The increase in enrollment from 1978 to 1979 was closest to

(A) 500        (B) 1000        (C) 3000

(D) 3500        (E) 5000

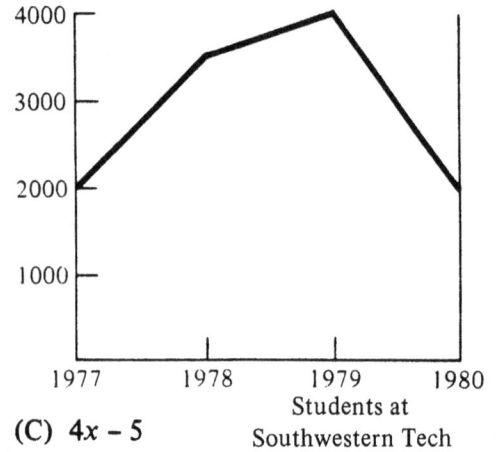

Students at
Southwestern Tech

21. If $x$ represents a number, express 5 less than 4 times the number.

(A) $5 - 4x$        (B) $5 + 4x$        (C) $4x - 5$

(D) $4(x - 5)$        (E) $\dfrac{x - 5}{4}$

22. Subtract $2x^2 - 3x + 1$ from $x^2 - 6$.

(A) $3x^2 - 3x - 5$        (B) $x^2 - 3x - 5$        (C) $x^2 - 3x + 7$

(D) $-x^2 + 3x - 7$        (E) $x^2 - 7$

23. $4ab(a^2 - 2ab^2) =$

(A) $4a^3b - 2ab^2$        (B) $4a^3b - 8a^2b^3$        (C) $2a^3b$

(D) $4a^2b - 2ab^2$        (E) $4a^3b - 8ab^2$

24. $3x^2(2x^3y^2)^2 =$

(A) $12x^8y^4$        (B) $6x^8y^4$        (C) $6x^5y^4$

(D) $12x^7y^4$        (E) $36x^8y^4$

25. $\dfrac{15x^4 - 25x^2}{5x} =$

(A) $3x^4 - 5x^2$        (B) $3x^3 - 5x$        (C) $3x^3 + 5x$

(D) $3x^3 - 5$        (E) $3x^2 - 25$

26. Factor $12a^3b - 9ab$.

(A) $12ab(4a^2 - 3)$        (B) $ab(12a^2 - 9)$        (C) $3ab(4a^2 - 3)$

(D) $12ab(3a^2 - 3)$        (E) $3(4a^3b - 3ab)$

27. Find $x$ if $4x - 7 = 2x + 9$.

(A) $\dfrac{9}{2}$        (B) 8        (C) $-8$        (D) $\dfrac{8}{3}$        (E) $-\dfrac{8}{3}$

| | A | B | C | D | E |
|---|---|---|---|---|---|
| 20 | ⫿ | ⫿ | ⫿ | ⫿ | ⫿ |
| 21 | ⫿ | ⫿ | ⫿ | ⫿ | ⫿ |
| 22 | ⫿ | ⫿ | ⫿ | ⫿ | ⫿ |
| 23 | ⫿ | ⫿ | ⫿ | ⫿ | ⫿ |
| 24 | ⫿ | ⫿ | ⫿ | ⫿ | ⫿ |
| 25 | ⫿ | ⫿ | ⫿ | ⫿ | ⫿ |
| 26 | ⫿ | ⫿ | ⫿ | ⫿ | ⫿ |
| 27 | ⫿ | ⫿ | ⫿ | ⫿ | ⫿ |

28. Find $t$ if $\frac{t+3}{4} = \frac{t+1}{3}$.

    (A) 2          (B) –3          (C) –1          (D) 0          (E) 5

29. If $\frac{3a - 2b + c}{4} = 1$, then $c =$

    (A) $4 - 3a + 2b$          (B) $1 - 3a + 2b$          (C) $4 + 3a - 2b$

    (D) $4(3a - 2b)$          (E) $4(1 - 3a + 2b)$

30. Which of these points lies on the graph of

    $$4x - 3y = 12$$

    (A) $(3, 4)$          (B) $(4, 3)$          (C) $(2, -1)$

    (D) $(6, 4)$          (E) $(4, 6)$

31. If the point $(0, b)$ lies on the graph of $5x - 2y = 4$, then $b =$

    (A) 2          (B) –2          (C) 4          (D) $\frac{4}{5}$          (E) $-\frac{2}{5}$

32. An equation for the line at the right is

    (A) $x = 2$                    (B) $y = -3$

    (C) $3x - 2y = 6$          (D) $3y - 2x = 6$

    (E) $2x - 3y = 6$

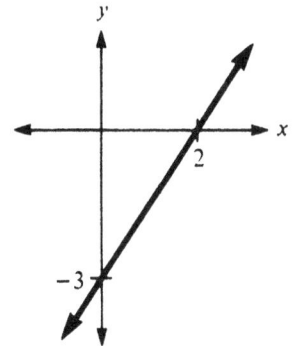

33. If $5x - y = 8$ and $3x - 2y = -5$, then

    (A) $x = 0, y = -8$          (B) $x = 2, y = 2$          (C) $x = 4, y = 12$

    (D) $x = 3, y = 7$          (E) $x = -1, y = 13$

34. Find the value of $3p^2 - 2pq$ when $p = -2$ and $q = 3$.

    (A) 48          (B) 24          (C) –36          (D) –24          (E) 60

35. Find the value of $\frac{3}{2^x}$ when $x = 3$.

    (A) 1          (B) $\frac{1}{2}$          (C) $\frac{3}{4}$          (D) $\frac{3}{8}$          (E) $\frac{27}{8}$

36. Find the value, in cents, of $n$ nickels, $q$ quarters, and 3 pennies.

    (A) $n + q + 3$          (B) $3nq$          (C) $3(n + q)$

    (D) $5n + 25q + 3$          (E) $5(n + q + 3)$

28  A B C D E
29  A B C D E
30  A B C D E
31  A B C D E
32  A B C D E
33  A B C D E
34  A B C D E
35  A B C D E
36  A B C D E

37. A psychologist needs 9 pounds of grain to feed 50 rats. If the rat population increases to 75, how much grain is needed to feed them?

(A) 6 pounds          (B) 12 pounds          (C) 13.5 pounds

(D) 15 pounds         (E) 24 pounds

38. If $\frac{x}{3} - 2 = \frac{x-2}{5}$, then $x =$

(A) 3          (B) 5          (C) 10          (D) 12          (E) 30

39. In the right triangle shown here, find $x$.

(A) 3          (B) $\sqrt{3}$          (C) 9

(D) $\sqrt{17}$          (E) $\sqrt{15}$

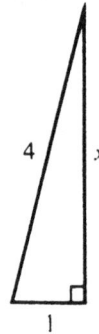

40. Out of 60 guidance counsellors at a college, 25% are women. Without firing anyone, the college wishes to bring the percentage of women up to 50 percent. How many women must it hire?

(A) 15          (B) 25          (C) 30          (D) 45          (E) 60

# Answers

1.     *i.*   53,808
     *ii.*   .0058

2.     *i.*   292
     *ii.*   122,385
     *iii.*   606
     *iv.*   (a) 21   (b) 23

3.     *i.*   80
     *ii.*   60

4.     *i.*   3 hours 33 minutes
     *ii.*   4 pounds 15 ounces
     *iii.*   8 feet 4 inches

5.     *i.*   $\frac{19}{20}$
     *ii.*   $\frac{17}{24}$

6.     *i.*   $\frac{1}{12}$
     *ii.*   $\frac{1}{6}$

7.     *i.*   $3\frac{5}{6}$
     *ii.*   $8\frac{1}{8}$
     *iii.*   2

8.     *i.*   (D) $\frac{2}{7}$

9.     *i.*   (E) .4001

10.    *i.*   24.985
     *ii.*   9.81

11.    *i.*   .264
     *ii.*   $388.05
     *iii.*   6000
     *iv.*   8

12.    *i.*   $21.82
     *ii.*   $2440
     *iii.*   $100

13.    *i.*   .18
     *ii.*   $\frac{3}{250}$
     *iii.*   $\frac{14}{25}$

14.    *i.*   24
     *ii.*   75
     *iii.*   60%
     *iv.*   100 tons

15.    *i.*   $1.26
     *ii.*   $18.75

16.    *i.*   $648
     *ii.*   26 feet
     *iii.*   50 inches

17.    *i.*   (D) 12,000
     *ii.*   (A) food
     *iii.*   (D) $250,000

18.    *i.*   13
     *ii.*   30
     *iii.*   -3

19.    *i.*   -5
     *ii.*   -17

20.    *i.*   $3x + 2$
     *ii.*   $5x + 10y$

21.    *i.*   52
     *ii.*   34

22.    *i.*   $11x + y$
     *ii.*   $8x^2 - 6x + 8$

23.    *i.*   $-20x^6y^4$
     *ii.*   $4p^3q - 8p^2q^2$

24.    *i.*   $100x^6y^8$

25.    *i.*   $5a + 6$

26.    *i.*   $5a^2b(3ab + 5)$

27.    *i.*   2

28.    *i.*   6
     *ii.*   20

29.    *i.*   $\frac{5a + b - 10}{2}$
     *ii.*   $\frac{5y - 2}{3}$

30.    *i.*   P: (a) 3   (b) 1   (c) P = (3, 1)
       Q: (a) -1   (b) 4   (c) Q = (-1, 4)
       R: (a) -3   (b) 0   (c) R = (-3, 0)
       S: (a) -4   (b) -4   (c) S = (-4, -4)
       T: (a) 2   (b) -3   (c) T = (2, -3)

31.    *i.*   (D) (4, 4)
     *ii.*   (C) $y = 6$

32.    *i.*   -3
     *ii.*   -2
     *iii.*   (B) $y = 4 - 2x$

33.    *i.*   (B) $x = 1, y = 1$
     *ii.*   (C) $x = 0, y = -1$

34.    *i.*   3100
     *ii.*   25
     *iii.*   240

35.    *i.*   $\sqrt{39}$
     *ii.*   $\sqrt{13}$

## Section 1 (page 12)

1.  245
3.  780
5.  7602
7.  6500
9.  63,056
11. 400,212
13. 747,008
15. 198,240
17. 2,001,096
19. 10,163,000
21. .4
23. .136
25. .3
27. .003
29. .038
31. .0308
33. 49.6
35. 6100.5
37. (A) 508
39. (D) 5,084,005
41. (C) .505
43. (B) 9,800,000
45. (E) 600.006
47. (E) .305
49. (E) .0063
51. (B) 555,493

**Section 2** (page 20)

| | | | |
|---|---|---|---|
| **1.** 958 | **3.** 1645 | **5.** 1660 | **7.** 1000 |
| **9.** 223 | **11.** 667 | **13.** 331 | **15.** 109 |
| **17.** 684 | **19.** 991 | **21.** 206 | **23.** 22,446 |
| **25.** 73,485 | **27.** 463,740 | **29.** 4,994,703 | **31.** 23,616,044 |
| **33.** 155,136 | **35.** 31 | **37.** 26 | **39.** 16 |
| **41.** 310 | **43.** 57 | **45.** 106 | **47.** 45 |
| **49.** (a) 6  (b) 11 | | **51.** (a) 22  (b) 8 | |
| **53.** (a) 15  (b) 12 | | **55.** (a) 114 (b) 10 | |
| **57.** $244 | | **59.** $27 | |
| **61.** 275 miles | | **63.** 15 | |

**Section 3** (page 25)

| | | | |
|---|---|---|---|
| **1.** 17 | **3.** 61 | **21.** 12 | **23.** 22 centimeters |
| **5.** 40 | **7.** 72 | **25.** (A) 22 | **27.** (C) 43 |
| **9.** 92 | **11.** 54.5 | **29.** (C) 44 | **31.** (D) 72 |
| **13.** 78.5 | **15.** 47 | **33.** (B) 79 | **35.** (C) 48.75 |
| **17.** 86 | **19.** 96 | | |

**Section 4** (page 30)

| | | |
|---|---|---|
| **1.** 8 hours 52 minutes | **3.** 8 hours 10 minutes | **5.** 12 hours 1 minute |
| **7.** 4 hours 20 minutes | **9.** 5 hours 30 minutes | **11.** 4 hours 52 minutes |
| **13.** 3 hours 39 minutes | **15.** 4 minutes 51 seconds | **17.** 2 hours 25 minutes |
| **19.** 1 hour 45 minutes | **21.** 1 hour 45 minutes | **23.** 12 pounds |
| **25.** 22 pounds 6 ounces | **27.** 17 kilograms 500 grams | **29.** 13 grams 27 centigrams |
| **31.** 3 pounds 2 ounces | **33.** 2 pounds 4 ounces | **35.** 3 pounds 14 ounces |
| **37.** 5 kilograms 225 grams | **39.** 3 grams 55 centigrams | **41.** 15 pounds 1 ounce |
| **43.** 4 pounds 11 ounces | **45.** 6 kilograms 500 grams | **47.** 9 feet |
| **49.** 11 feet 5 inches | **51.** 10 kilometers | **53.** 3 feet 11 inches |
| **55.** 4 feet 3 inches | **57.** 2 kilometers 955 meters | **59.** 89 centimeters |

**Section 5** (page 39)

| | | | |
|---|---|---|---|
| **1.** $\frac{1}{2}$ | **3.** $\frac{5}{9}$ | **5.** $\frac{2}{3}$ | **7.** $\frac{3}{4}$ |
| **9.** $\frac{5}{7}$ | **11.** 1 | **13.** $\frac{1}{3}$ | **15.** $\frac{7}{10}$ |
| **17.** $\frac{1}{2}$ | **19.** 35 | **21.** 24 | **23.** 90 |
| **25.** 576 | | | |

**27.** (a) 4    (b) $\frac{1}{2} = \frac{2}{4}, \frac{1}{4} = \frac{1}{4}$    **29.** (a) 12    (b) $\frac{3}{4} = \frac{9}{12}, \frac{5}{6} = \frac{10}{12}$

**31.** (a) 36    (b) $\frac{2}{9} = \frac{8}{36}, \frac{1}{12} = \frac{3}{36}$    **33.** (a) 48    (b) $\frac{7}{12} = \frac{28}{48}, \frac{3}{16} = \frac{9}{48}$

**35.** (a) 120    (b) $\frac{1}{30} = \frac{4}{120}, \frac{7}{40} = \frac{21}{120}$    **37.** $\frac{5}{6}$

39. $\frac{4}{15}$        41. $\frac{25}{63}$        43. $\frac{37}{40}$        45. $\frac{29}{60}$

47. $\frac{7}{24}$        49. $\frac{3}{50}$        51. $\frac{25}{36}$        53. (E) $\frac{11}{12}$

55. (A) $\frac{1}{20}$    57. (C) $\frac{5}{6}$    59. (A) $\frac{37}{40}$    61. (E) $\frac{19}{24}$

63. (A) $\frac{23}{42}$   65. (C) $\frac{29}{24}$   67. (C) $\frac{3}{8}$    69. (E) 70 centimeters

## Section 6 (page 45)

1. $\frac{1}{6}$          3. $\frac{1}{5}$          25. $\frac{10}{3}$         27. $\frac{5}{2}$

5. $\frac{2}{7}$          7. $\frac{1}{6}$          29. $\frac{1}{54}$         31. $\frac{1}{60}$

9. $\frac{2}{3}$          11. $\frac{5}{14}$        33. (B) $\frac{5}{28}$     35. (A) $\frac{3}{7}$

13. $\frac{1}{36}$        15. $\frac{1}{30}$        37. (B) $\frac{9}{10}$     39. (B) 10

17. $\frac{2}{3}$         19. $\frac{1}{2}$         41. (A) $\frac{5}{32}$     43. (B) $\frac{27}{2}$

21. $\frac{5}{12}$        23. $\frac{3}{4}$         45. (D) $\frac{1}{8}$

## Section 7 (page 53)

1. $\frac{3}{2}$          3. $\frac{21}{5}$         33. $3\frac{1}{6}$         35. $3\frac{5}{13}$

5. $\frac{27}{5}$         7. $\frac{23}{3}$         37. 2                      39. $\frac{91}{8}$

9. $5\frac{1}{2}$         11. 6                     41. $\frac{20}{17}$        43. (B) $7\frac{1}{4}$

13. $7\frac{3}{4}$        15. $3\frac{9}{10}$       45. (D) $7\frac{6}{7}$     47. (D) $\frac{81}{8}$

17. $9\frac{1}{8}$        19. $1\frac{3}{4}$        49. (B) $3\frac{1}{4}$     51. (C) $2\frac{1}{2}$

21. $2\frac{1}{2}$        23. $\frac{3}{8}$         53. (D) $2\frac{9}{10}$    55. (D) $\frac{45}{28}$

25. 10                    27. $7\frac{7}{8}$        57. (D) $\frac{33}{20}$    59. (D) $2\frac{5}{8}$ yards

29. $4\frac{7}{32}$       31. 4

## Section 8 (page 60)

1. (A) $\frac{2}{5}$      3. (C) $\frac{1}{10}$     29. (E) $\frac{8}{7}$      31. (A) $\frac{7}{10}$

5. (D) $\frac{2}{15}$     7. $\frac{1}{2}$          33. (A) $\frac{8}{11}$     35. (E) $\frac{4}{15}$

9. $\frac{1}{3}$          11. $\frac{5}{6}$         37. (A) $\frac{1}{2}$      39. (E) $\frac{7}{10}$

13. $\frac{9}{11}$        15. $\frac{7}{13}$        41. (D) $\frac{13}{16}$    43. (A) $\frac{5}{3}$

17. (C) $\frac{1}{5}$     19. (A) $\frac{1}{10}$    45. $\frac{1}{2}$ pound of feathers

21. (E) $\frac{2}{11}$    23. (C) $\frac{2}{5}$     47. The side that measures $\frac{7}{10}$ of a meter

25. (D) $\frac{1}{10}$    27. (B) $\frac{9}{100}$   49. The 8-ounce bar

**Section 9** (page 68)

| | | | |
|---|---|---|---|
| 1.  (E) .1 | 3.  (D) .11 | 25.  (E) .1501 | 27.  (D) .07 |
| 5.  (D) .38 | 7.  (A) .09 | 29.  (E) 10.009 | 31.  (B) 3.389 |
| 9.  (C) .39 | 11.  (A) .451 | 33.  (A) .004 | 35.  (C) .0039 |
| 13.  (B) .575 | 15.  (D) .100 | 37.  (D) .0202 | 39.  (C) .87 |
| 17.  (E) .136 | 19.  (B) .703 | 41.  (E) 2.11 | 43.  (A) 6.1 |
| 21.  (E) .601 | 23.  (E) .7 | 45.  (E) 6.0101 | |

**Section 10** (page 71)

| | | | |
|---|---|---|---|
| 1.  20.5 | 3.  14.73 | 25.  24.51 | 27.  44.77 |
| 5.  26.61 | 7.  39.828 | 29.  11.77 | 31.  33.33 |
| 9.  115.44 | 11.  133.211 | 33.  27.111 | 35.  117.483 |
| 13.  121.989 | 15.  1.4 | 37.  5.91 | 39.  16.14 |
| 17.  10.9 | 19.  6.04 | 41.  15.764 | |
| 21.  41.73 | 23.  .183 | | |

**Section 11** (page 76)

| | | | |
|---|---|---|---|
| 1.  .15 | 3.  .112 | 29.  40,000 | 31.  20 |
| 5.  .08 | 7.  .2668 | 33.  3 | 35.  40 |
| 9.  2.4 | 11.  114.8 | 37.  18 | 39.  24 |
| 13.  1081.45 | 15.  $7.80 | 41.  15 | 43.  72 |
| 17.  $61.50 | 19.  $3.40 | 45.  (E) .0108 | 47.  (D) 8.841 |
| 21.  1.5 inches | 23.  $420.65 | 49.  (D) 40 | 51.  (A) $14.40 |
| 25.  50 | 27.  800 | 53.  (B) $29.70 | |

**Section 12** (page 80)

| | | | |
|---|---|---|---|
| 1.  (E) $44.16 | 3.  (A) $596 | 13.  (D) $1.90 | 15.  (C) $1520 |
| 5.  (D) $620 | 7.  $950 | 17.  $1315 | 19.  $588 |
| 9.  $4.00 | 11.  (C) $3.60 | 21.  $158.75 | |

**Section 13** (page 85)

| | | | |
|---|---|---|---|
| 1.  .8 | 3.  2.5 | 33.  $\frac{3}{4}$ | 35.  $\frac{11}{50}$ |
| 5.  .35 | 7.  .28 | 37.  $\frac{3}{2}$ | 39.  $\frac{12}{25}$ |
| 9.  .005 | 11.  .67 | 41.  .53 | 43.  .4 |
| 13.  .17 | 15.  .31 | 45.  .02 | 47.  2.25 |
| 17.  .455 | 19.  .615 | 49.  (C) $\frac{3}{20}$ | 51.  (D) .125 |
| 21.  .923 | 23.  $\frac{6}{25}$ | 53.  (C) .29 | 55.  (E) .06 |
| 25.  $\frac{1}{50}$ | 27.  $\frac{5}{8}$ | 57.  (C) .16 | 59.  (C) .417 |
| 29.  $\frac{6}{5}$ | 31.  $\frac{1}{2}$ | | |

**Section 14** (page 93)

| | | | |
|---|---|---|---|
| 1. (B) 33 | 3. (D) 45 | 33. (E) 144 | 35. 100 |
| 5. (D) 27 | 7. 288 | 37. 480 | 39. 10 |
| 9. 28 | 11. 68 | 41. (A) 50% | 43. (D) 80% |
| 13. 10 | 15. 7 | 45. 90% | 47. 10% |
| 17. 8 | 19. 7 | 49. (C) 41 | 51. (B) 40% |
| 21. 18 | 23. 90 | 53. (D) 600 | 55. 20 |
| 25. 20 | 27. 300 | 57. $8360 | 59. 6600 |
| 29. (D) 60 | 31. (C) 40 | | |

**Section 15** (page 98)

| | | | |
|---|---|---|---|
| 1. (B) $224 | 3. (C) $149.80 | 13. (E) $19,600 | 15. (D) $17.00 |
| 5. (D) $1.30 | 7. $47,840 | 17. (C) $22.20 | 19. 174,600 |
| 9. $5.04 | 11. $28,196 | 21. $14,080 | |

**Section 16** (page 107)

| | | | |
|---|---|---|---|
| 1. 16 | 3. 49 | 29. $210 | 31. $9 |
| 5. 144 | 7. (C) 18 ft.² | 33. (B) 26 in. | 35. (C) 22 ft. |
| 9. (D) 700 in.² | 11. 54 in.² | 37. 308 yd. | 39. (E) 4 in. |
| 13. 220 in.² | 15. 300 in.² | 41. 6 in. | 43. 24 in.² |
| 17. 225 ft.² | 19. (A) 10 yd. | 45. (C) $300 | 47. (D) $1400 |
| 21. 3 ft. | 23. (D) $450 | 49. $25 | 51. $420 |
| 25. (D) $180 | 27. $84 | | |

**Section 17** (page 113)

| | | | |
|---|---|---|---|
| 1. (B) 2500 | 3. (D) 4500 | 21. (C) $900,000 | 23. (D) $1,584,000 |
| 5. (D) 1979 | 7. (D) 140,000 | 25. (C) $288,000 | 27. (E) $100,000 |
| 9. (E) 20,000 | 11. (D) $5,000,000 | 29. (B) $200,000 | 31. (B) 2,750,000 |
| 13. (C) $9,000,000 | 15. (D) $1,000,000 | 33. (C) 1,250,000 | 35. (B) 500,000 |
| 17. (B) $325 | 19. (D) $1250 | | |

**Section 18** (page 127)

| | | | |
|---|---|---|---|
| 1. 5 | 3. 15 | 33. 23 | 35. 0 |
| 5. .73 | 7. $\frac{1}{3}$ | 37. -12 | 39. 12 |
| 9. -5 | 11. 2 | 41. 35 | 43. 36 |
| 13. 4 | 15. -20 | 45. 0 | 47. 96 |
| 17. -4 | 19. -3 | 49. -2 | 51. 2 |
| 21. 1 | 23. 12 | 53. -5 | 55. -4 |
| 25. -1 | 27. 7 | 57. undefined | 59. undefined |
| 29. -7 | 31. -7 | 61. undefined | 63. −16° Fahrenheit |
| | | 65. 3 miles west | |

**Section 19** (page 132)

| | | | |
|---|---|---|---|
| 1. 4 | 3. 36 | 13. 64 | 15. 1 |
| 5. 100 | 7. 121 | 17. (B) 20 | 19. (D) -20 |
| 9. 0 | 11. 27 | 21. 49 | 23. -19 |

| 25. (C) 42 | 27. (A) 50 | 41. -70 | 43. -14 |
| 29. (B) -199 | 31. 22 | 45. $100 | 47. 6 hours |
| 33. 14 | 35. 75 | 49. $49 | 51. 4 ·yards |
| 37. 26 | 39. -36 | | |

## Exam A on Arithmetic (page 135)

1.  (C) 220,500   2.  (B) 318   3.  (C) 312   4.  66

5.  (A) $1130   6.  (D) $\frac{4}{7}$   7.  6 hours 45 minutes

8.  $\frac{23}{45}$   9.  $1\frac{7}{10}$   10.  $1\frac{6}{7}$   11.  26.277

12.  (D) .0079   13.  14.65   14.  (B) .77   15.  $35.40

16.  45   17.  80   18.  $60   19.  (C) 7500

20.  1

## Exam B on Arithmetic (page 136)

1.  (C) 5000.2   2.  53   3.  (B) 89   4.  (C) 2 hours 45 minutes

5.  $\frac{7}{10}$   6.  $3\frac{5}{8}$   7.  (E) $\frac{3}{13}$   8.  (E) .004 49

9.  $3\frac{1}{13}$   10.  (C) 2.16   11.  (C) 150   12.  22.95

13.  (C) $\frac{22}{25}$   14.  $300   15.  3000   16.  10

17.  (C) $12   18.  (D) 450 square feet   19.  (A) $1,920,000

20.  13

## Section 20 (page 143)

1.  $x + 3$   3.  $6x$   5.  $x + 2$

7.  $x + 12$   9.  $x^2$   11.  $1 + 3x$ or $3x + 1$

13.  $\frac{1}{2}x - 6$ or $\frac{x}{2} - 6$   15.  (D) $5n + 10d$   17.  (E) $25q + 50h$

19.  $5x + 10y$   21.  $5x + 20y + 50z$   23.  $25x$

25.  (A) $50h$   27.  $mh$   29.  $80x + 60y$   31.  $20x + 17y$

## Section 21 (page 149)

| 1. (B) 12 | 3. (C) 0 | 25. (C) -4 | 27. -1 |
| 5. (D) -15 | 7. 16 | 29. 33 | 31. 77 |
| 9. -9 | 11. 10,000 | 33. -1 | 35. (B) 24 |
| 13. 12 | 15. -8 | 37. (E) 0 | 39. -23 |
| 17. (D) 16 | 19. 35 | 41. 25 | 43. 33 |
| 21. 980 | 23. (D) -2 | 45. 24 | 47. -6 |

49.  (a)  49 square inches   (b)  144 square inches

51.  (a)  1200 miles   (b)  3600 miles

53.  (a)  460 square feet   (b)  684 square feet

55.  (a)  $28   (b)  $49

57.  (a)  $3.75

**Section 22** (page 158)

| | | | |
|---|---|---|---|
| 1. | 20 | 3. | 1 |
| 5. | 16 | 7. | $x^2$ and $4x$ |
| 9. | $3b^2$ and $-5$ | 11. | unlike terms |
| 13. | like terms | 15. | like terms |
| 17. | unlike terms | 19. | like terms |
| 21. | $12a$ | 23. | $x$ |
| 25. | 0 | 27. | $12xy$ |

| | | | |
|---|---|---|---|
| 29. | (A) $6x + 10y$ | 31. | (D) $7x$ |
| 33. | $a - 1$ | 35. | $8x + 2y$ |
| 37. | $5x - y$ | 39. | $6x^2 + 2x + 3$ |
| 41. | (E) $5x + 3y$ | 43. | (D) $4p + 9q$ |
| 45. | $3x^2 - 2x + 7$ | 47. | $-4a + 2b + 3c$ |
| 49. | $x^2 - 7x - 1$ | 51. | (D) $10x - 2$ |
| 53. | (E)  $3x - 1$ | | |

**Section 23** (page 164)

| | | | |
|---|---|---|---|
| 1. | (B) $a^7$ | 3. | (D) $x^4$ |
| 5. | $m^8$ | 7. | $t^{11}$ |
| 9. | (B) $4a^3b$ | 11. | (A) $27x^5y^2$ |
| 13. | $-10m^5n^2$ | 15. | $21p^6q^6$ |
| 17. | $6a^2b^3c$ | 19. | (D) $-15a^4bc^8$ |
| 21. | (E) $32x^5y^3z^{10}$ | 23. | (B) $a^2 + 3a$ |

| | | | |
|---|---|---|---|
| 25. | $3c^3 + 12c$ | 27. | $4y^3 - 3y^2$ |
| 29. | $8t^3 - 12t^2$ | 31. | $6x^3y - 3x^2y$ |
| 33. | $-4x^3y^3 + 4x^2y^4$ | 35. | (B) $x^2 + 3x$ |
| 37. | (B) $4x^2 - 2x$ | 39. | (E) $x$ |
| 41. | $10x^2y^2 - 4x^3y$ | 43. | $8xy - 2x^2y$ |
| 45. | $16x^2y^2 - 3x^2y$ | | |

**Section 24** (page 171)

| | | | |
|---|---|---|---|
| 1. | $s^8$ | 3. | $u^{12}$ |
| 5. | $y^{10}$ | 7. | (D) $a^{15}$ |
| 9. | (A) $c^{21}$ | 11. | (E) $y^{30}$ |
| 13. | (A) $a^2b^4$ | 15. | (C) $p^{12}q^4$ |
| 17. | (E) $4a^6$ | 19. | $25y^6$ |

| | | | |
|---|---|---|---|
| 21. | $1000u^{15}$ | 23. | $27x^6$ |
| 25. | $4x^2y^6$ | 27. | $25p^2q^8$ |
| 29. | $100x^4y^8$ | 31. | (D) $a^5b^2$ |
| 33. | (E) $100x^8$ | 35. | $25x^8y^2$ |
| 37. | $27x^8y^6$ | 39. | $4a^5b^7$ |

**Section 25** (page 176)

| | | | |
|---|---|---|---|
| 1. | $a$ | 3. | $3c$ |
| 5. | $3x^2$ | 7. | $-2m^2$ |
| 9. | $x$ | 11. | $2s^3$ |
| 13. | $-6x^3$ | 15. | 1 |
| 17. | 2 | 19. | $c$ |
| 21. | $5x^2$ | 23. | $-8s^4$ |

| | | | |
|---|---|---|---|
| 25. | (D) $a + 1$ | 27. | (D) $2t + 2$ |
| 29. | $z - 3$ | 31. | $3b - 5$ |
| 33. | (C) $a + 1$ | 35. | $c^2 + 1$ |
| 37. | $2x + 5$ | 39. | (C) $x + 1$ |
| 41. | (B) $2x + 3$ | 43. | $2x - 3$ |
| 45. | $3b^2 - 4$ | 47. | $3x^3 + 4x$ |

**Section 26** (page 182)

| | | | |
|---|---|---|---|
| 1. | 4 | 3. | 6 |
| 5. | 2 | 7. | 3 |
| 9. | 5 | 11. | 3 |
| 13. | 1 | 15. | 4 |
| 17. | 3 | 19. | 5 |
| 21. | 6 | 23. | (A) 3 |

| | | | |
|---|---|---|---|
| 25. | (E) 4 | 27. | (C) $b$ |
| 29. | (D) $3y$ | 31. | (B) 4 |
| 33. | (C) $3x$ | 35. | $4a$ |
| 37. | (B) $6a(a + 1)$ | 39. | (C) $4x(x + 2)$ |
| 41. | $10x(2x^2 + 3)$ | 43. | $2xy(x + 2)$ |
| 45. | $6pq(2p - 3q)$ | 47. | $a(9ab^2 + 8)$ |

**Section 27** (page 190)

| | | | |
|---|---|---|---|
| 1. | yes | 3. | no |
| 5. | no | 7. | no |
| 9. | no | 11. | no |
| 13. | 10 | 15. | 5 |

| | | | |
|---|---|---|---|
| 17. | 12 | 19. | $-18$ |
| 21. | $-4$ | 23. | $-8$ |
| 25. | 7 | 27. | $-6$ |
| 29. | $-11$ | 31. | $-6$ |

| | | | |
|---|---|---|---|
| **33.** -1 | **35.** 5 | **49.** 2 | **51.** -25 |
| **37.** (D) 3 | **39.** (A) 6 | **53.** $\frac{5}{2}$ | **55.** 8 |
| **41.** 5 | **43.** 2 | | |
| **45.** (C) 1 | **47.** (D) 2 | **57.** $-\frac{3}{5}$ | |

**Section 28** (page 196)

| | | | |
|---|---|---|---|
| **1.** (C) 10 | **3.** (E) 49 | **25.** 5 | **27.** 2 |
| **5.** (C) 6 | **7.** 10 | **29.** 4 | **31.** -1 |
| **9.** $\frac{3}{5}$ | **11.** $\frac{3}{2}$ | **33.** $\frac{1}{2}$ | **35.** (A) 6 |
| **13.** 9 | **15.** 6 | **37.** (E) 20 | **39.** 10 |
| **17.** 10 | **19.** (D) 6 | **41.** -24 | **43.** 36 |
| **21.** 30 | **23.** (D) 5 | **45.** 15 | **47.** -6 |

**Section 29** (page 203)

| | | | |
|---|---|---|---|
| **1.** $y + 8$ | **3.** $\frac{y}{4}$ | **5.** $\frac{a}{3}$ | **7.** $\frac{y+1}{2}$ |
| **9.** $\frac{1+3n}{5}$ | **11.** $7a - 2$ | **13.** $1 - 2y$ | **15.** $\frac{5c-b}{a}$ |
| **17.** $3t + s$ | **19.** $bc - 1$ | **21.** (C) $\frac{1-y}{2}$ | **23.** (A) $4t - 1$ |
| **25.** (B) $\frac{1-2x}{3}$ | **27.** (C) $\frac{10-2s}{5}$ | **29.** (A) $\frac{y+3}{2}$ | **31.** (B) $\frac{c-2b}{a}$ |
| **33.** (C) $\frac{1+y-5z}{3}$ | **35.** (D) $\frac{4y+1}{3}$ | | |

**37.** (a) $l = \dfrac{A}{w}$ (b) 7 meters

**39.** (a) $b = \dfrac{2A}{b}$ (b) 30 inches

**41.** (a) $r = \dfrac{d}{t}$ (b) 120 kilometers per hour

**43.** (a) $P = \dfrac{I}{R \cdot T}$ (b) \$20,000

**Section 30** (page 210)

**1.–8.**   **9.–16.**

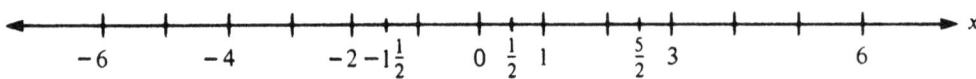

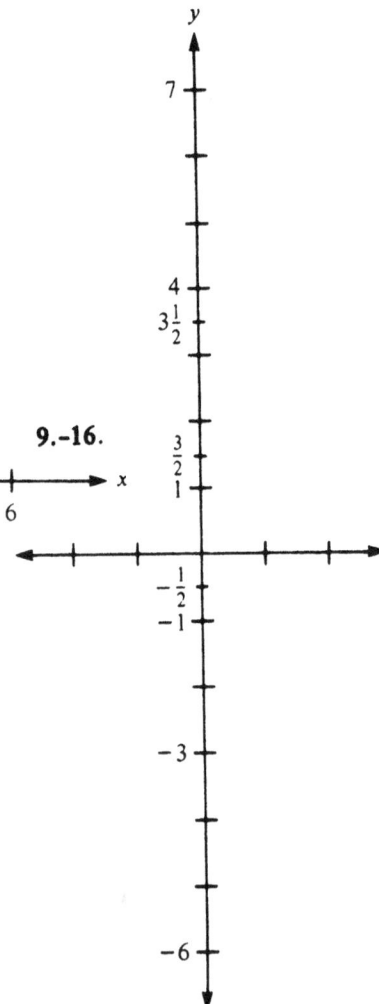

| | | | |
|---|---|---|---|
| **17.** (a) 2 | (b) 2 | (c) $P = (2, 2)$ |
| **19.** (a) 3 | (b) 5 | (c) $R = (3, 5)$ |
| **21.** (a) -2 | (b) 2 | (c) $T = (-2, 2)$ |
| **23.** (a) -5 | (b) 0 | (c) $V = (-5, 0)$ |
| **25.** (a) 0 | (b) -3 | (c) $X = (0, -3)$ |
| **27.** (a) $\frac{1}{2}$ | (b) 0 | (c) $Z = \left(\frac{1}{2}, 0\right)$ |

**28.-38.**

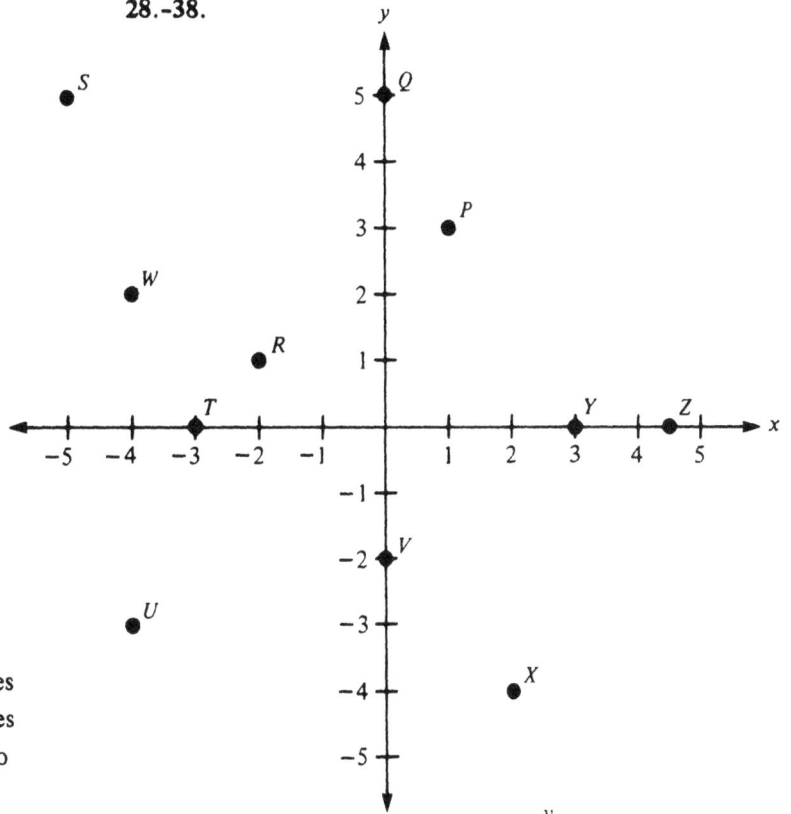

**Section 31** (page 218)

| | | | |
|---|---|---|---|
| **1.** | yes | **3.** | yes |
| **5.** | no | **7.** | yes |
| **9.** | no | **11.** | no |

**13.**

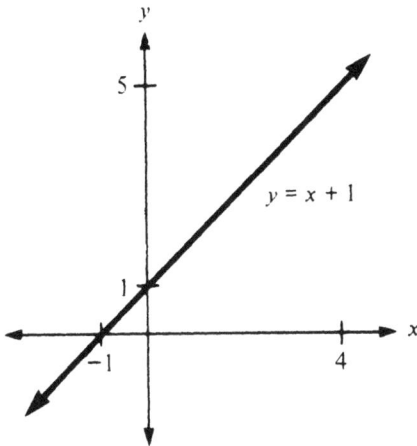

$y = x + 1$

**15.**

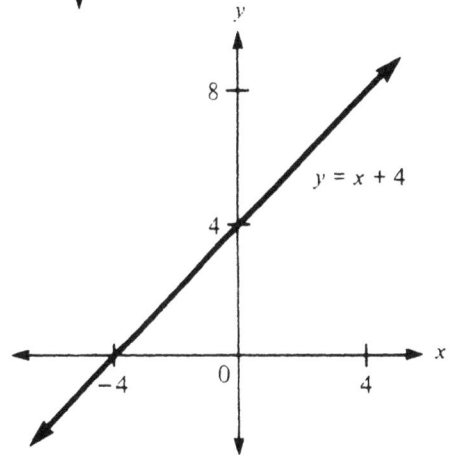

$y = x + 4$

**17.**

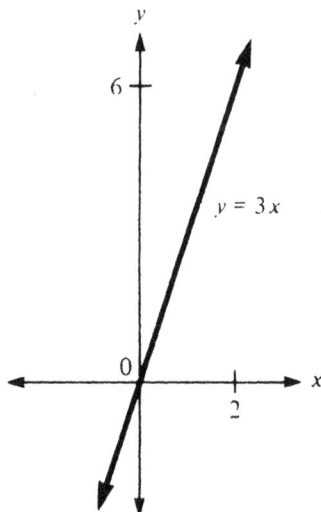

$y = 3x$

**19.**

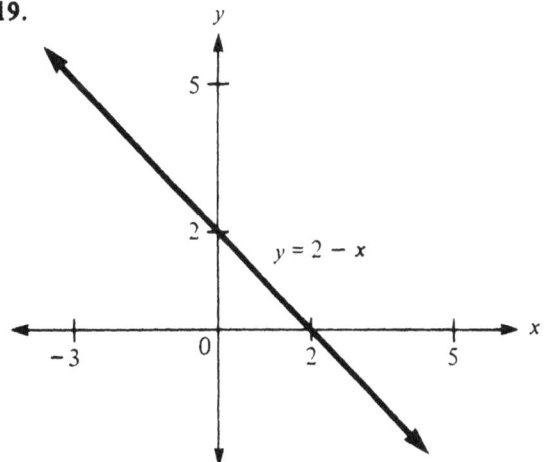

$y = 2 - x$

**21.**

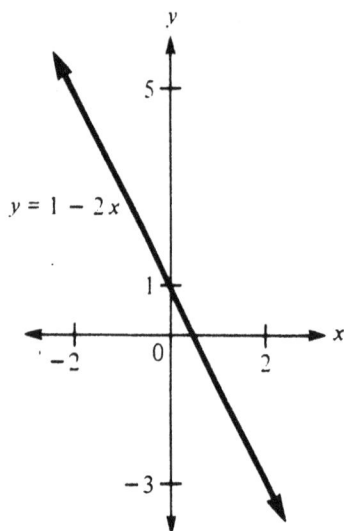

$y = 1 - 2x$

**23.** (C) $(2, 4)$

**25.** (E) $(0, 5)$

**27.** (A) $(0, 3)$

**29.** (B) $(2, 2)$

**31.** (E) $(6, -4)$

**33.** (D) $\left(2, -\frac{1}{3}\right)$

**35.** neither

**37.** (b) vertical

**39.** (b) vertical

**41.** neither

**43.**

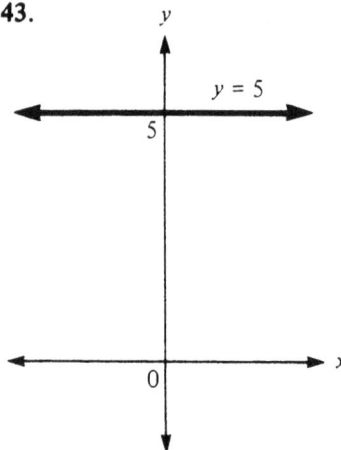

$y = 5$

**45.** (D) $(0, -3)$

## Section 32 (page 228)

**1.** (B) $(0, 4)$

**3.** (D) $(0, 0)$

**5.** (E) $(0, -5)$

**7.** $(0, -5)$

**9.** $(0, 6)$

**11.** $(6, 0)$

**13.** $(0, -3)$

**15.** (A) $6$

**17.** (B) $-8$

**19.** (B) $-5$

**21.** $6$

**23.** $2$

**25.** $4$

**27.** $7$

**29.** (D) $x + y = 3$

**31.** (A) $y = 4$

**33.** (E) $y = x - 2$

**35.** (D) $2y - x = 4$

**37.** (E) $5y = x - 5$

**39.** (C) $y - x = 2$

## Section 33 (page 239)

**1.** yes

**3.** no

**5.** yes

**7.** yes

**9.** yes

**11.** $(4, 2)$

**13.** $(4, 6)$

**15.** $(4, 1)$

**17.** $(1, 2)$

**19.** (E) $x = 7, y = 5$

**21.** (B) $x = 3, y = 7$

**23.** (A) $x = 4, y = 1$

**25.** (B) $(5, 1)$

**27.** $(1, -1)$

**29.** (A) $x = 2, y = 3$

**31.** $x = 2, y = 1$

**33.** (D) $x = 6, y = 9$

**35.** (D) $x = 4, y = 3$

**37.** $x = 1, y = 2$

**39.** $x = 5, y = 2$

## Section 34 (page 248)

**1.** (D) $900$

**3.** (B) $375$

**5.** 8 meters

**7.** $4000$

**9.** (C) 15 inches

**11.** (C) $14$

**13.** $50$

**15.** $198$

**17.** (B) $200$

**19.** $240$

**21.** (D) $65$

**23.** $\$260$

**25.** $79$

**27.** $93$

**29.** (C) $12$

**31.** (C) $20$

## Section 35 (page 258)

**1.** $4$

**3.** $8$

**5.** $7$

**7.** $11$

**9.** $0$

**11.** (B) $20$

**13.** (C) $90$

**15.** $800$

**17.** (D) $6$

**19.** $5$

**21.** $8$

**23.** (B) $\frac{1}{4}$

25. $\frac{2}{9}$    27. $\frac{5}{8}$    41. $\sqrt{41}$    43. (A) 10

29. $\frac{10}{9}$    31. (D) 3    45. (D) 12    47. (D) 7

33. (B) 3    35. $\sqrt{13}$    49. (E) $\sqrt{125}$    51. (A) 2

37. (C) 5    39. (D) 6

**Exam A on Algebra** (page 262)

1.  $7x - y$    2.  $-20a^4bc^4$    3.  $x^2 - 8x$    4.  $6a^3 + 9$

5.  (D) $27x^6y^9$    6.  (B) $2a + 10$    7.  $9a(a - 2)$    8.  5

9.  (D) $25q + 10d$    10.  33    11.  $\frac{35}{2}$    12.  $-12$

13.  (D) $\frac{a + 3}{5}$    14.  (C) $(3, 1)$    15.  (C) $-7$    16.  (D) $2x - y = 2$

17.  (B) 80    18.  (C) 9    19.  (E) $x = 3, y = -1$  20.  (C) 2000

**Exam B on Algebra** (page 264)

1.  (C) $8x - y$    2.  $-8x^4y^4$    3.  (B) $a^2b + 4ab^2$    4.  $6x^2 - 4x - 6$

5.  (D) $6a - 1$    6.  (D) $64p^6q^3$    7.  (E) $5n + 25q$    8.  $3ab(2a - 9)$

9.  (C) 5    10.  (C) 12    11.  (C) 8    12.  (C) $(2, -2)$

13.  44    14.  (D) $y = x + 2$    15.  (B) $\frac{y + 9}{5}$    16.  (C) \$105

17.  (A) 10    18.  (E) $x = 1, y = 2$    19.  (A) 87    20.  (D) 75

# Complete Solutions to the Sample Final Exams

**Sample Final Exam A** (page 266)

1. (D) 420,500    2. (A) 249    3. (B) 407    4. (C) 83

$$\begin{array}{r} 407 \\ 16\overline{)6512} \\ \underline{64\phantom{00}} \\ 112 \\ \underline{112} \end{array}$$

$$\frac{78 + 81 + 82 + 91}{4} = \frac{332}{4} = 83$$

5. (B) \$2707

PROFIT = SALES − COSTS
= 952 × \$3 − (\$97 + \$52)
= \$2856 − \$149
= \$2707

6. (B) $\frac{3}{5}$

7. (D) 8 pounds 14 ounces

$$\begin{array}{l} \phantom{-}26 \text{ pounds} \quad 28 \text{ ounces} \\ \cancel{27 \text{ pounds } 12 \text{ ounces}} \\ -18 \text{ pounds } 14 \text{ ounces} \\ \hline \phantom{-18 \text{ pounds } }8 \text{ pounds } 14 \text{ ounces} \end{array}$$

8. (C) $\frac{35}{36}$    The *lcd* is 36.    $\frac{2}{9} = \frac{8}{36}$,    $\frac{3}{4} = \frac{27}{36}$    $\frac{2}{9} + \frac{3}{4} = \frac{8}{36} + \frac{27}{36} = \frac{35}{36}$

9. (B) $2\frac{1}{6}$    $4\frac{1}{2} = 4\frac{3}{6}$,    $2\frac{1}{3} = 2\frac{2}{6}$

$$\begin{array}{r} 4\frac{3}{6} \\ -2\frac{2}{6} \\ \hline 2\frac{1}{6} \end{array}$$

10. (A) 3    $8\frac{1}{4} = \frac{33}{4}$,    $2\frac{3}{4} = \frac{11}{4}$

$$8\frac{1}{4} \div 2\frac{3}{4} = \frac{33}{4} \div \frac{11}{4} = \frac{\overset{3}{\cancel{33}}}{\cancel{4}} \times \frac{\cancel{4}}{\cancel{11}}^{1} = 3$$

11. (B) 26.072

$$\begin{array}{r} 15.093 \\ +\phantom{0}2.979 \\ +\phantom{0}8.\phantom{000} \\ \hline 26.072 \end{array}$$

12. (C) .0309

13. (B) 79.84

$$\begin{array}{r} 87.80 \\ -\phantom{0}7.96 \\ \hline 79.84 \end{array}$$

14. (A) .92

$$\begin{array}{r} .923\phantom{0} \\ 13\overline{)12.000} \\ \underline{11\,7\phantom{00}} \\ 30\phantom{0} \\ \underline{26\phantom{0}} \\ 40 \\ \underline{39} \end{array}$$

To the nearest hundredth,

$$\frac{12}{13} \approx .92$$

15. (C) \$127.20

$$\begin{array}{r} \$\phantom{0}7.95 \longleftarrow \text{2 decimal digits} \\ \times\phantom{00}16 \longleftarrow + \text{0 decimal digits} \\ \hline 47\,70\phantom{0} \\ \underline{79\,5\phantom{00}} \\ \$127.20 \longleftarrow \text{2 decimal digits} \end{array}$$

16. (D) 45    60% *of* 75 = .60 × 75 = 45.00 = 45

**17. (D) 75**

Let $x$ be the number in question.

60 is 80% of a certain number.

$60 = .8\emptyset \times \quad\quad x$    Multiply both sides by 10.

$600 = 8x$

$\dfrac{600}{8} = x$

$75 = x$

**18. (D) $40.50**

A 10% reduction on a $45 dress means $.10 \times \$45$, which equals $4.50.

$\$45.00$ ←— original price

$-\quad 4.50$ ←— reduction

$\$40.50$ ←— sale price

**19. (B) 880 in.$^2$**

Area = length $\times$ width

$= 40$ in. $\times 22$ in.

$= 880$ in.$^2$

**20. (C) 12,500**

3 000 ←— September

2 000 ←— October

3 500 ←— November

$\underline{4\ 000}$ ←— December

12,500

**21. (A) 10x**

$4x + 9y$

$\underline{6x - 9y}$

$10x$

**22. (B) $-12a^3b^5c^5$**

$(6a^2bc^3)(-2ab^4c^2) = \big(6 \times (-2)\big)(a^2 \cdot a)(b \cdot b^4)(c^3 \cdot c^2)$

$= -12a^{2+1}b^{1+4}c^{3+2}$

$= -12a^3b^5c^5$

**23. (D) $a^2 + 12a$**

$5a^2 - 4a(a - 3) = 5a^2 - 4a \cdot a - 4a(-3)$

$= 5a^2 - 4a^2 + 12a$

$= a^2 + 12a$

**24. (A) $4x^2 - 2x - 1$**

$\quad\ 3x^2 - 2x + 4$

$\underline{+\ \ x^2 \quad\quad\ -\ 5}$

$\quad\ 4x^2 - 2x - 1$

**25. (A) $25x^6y^8$**

$(5x^3y^4)^2 = 5^2(x^3)^2(y^4)^2$

$= 25x^{3\times2}y^{4\times2}$

$= 25x^6y^8$

**26. (D) $-3a - 4$**

$\dfrac{12a^2 + 16a}{-4a} = \dfrac{12a^2}{-4a} + \dfrac{16a}{-4a} = -3a - 4$

**27. (E) $3x^2(3x - 5)$**

**28. (B) 22**    $2(-5)^2 + 4(-7) = 2(25) + (-28) = 50 - 28 = 22$

**29. (D) 4**

$4x + 3 = 5x - 1$

$4x - 4x + 3 = 5x - 4x - 1$

$3 = x - 1$

$4 = x$

**30. (C) $5f + 10t$**

Each ten-dollar bill is worth 10 dollars;

$\quad$ $t$ ten-dollar bills are worth $t \times 10$ dollars or $10t$ dollars.

Each five-dollar bill is worth 5 dollars;

$\quad$ $f$ five-dollar bills are worth $f \times 5$ dollars or $5f$ dollars.

The total value, in dollars, is $10t + 5f$ or $5f + 10t$.

**31. (A) 76**

Substitute 4 for $x$ and $-3$ for $y$ in the expression $x^2 - 5xy$.

$\quad 4^2 - 5(4)(-3) = 16 - 20(-3) = 16 + 60 = 76$

**32. (C) $-\dfrac{15}{4}$**

$\dfrac{x}{5} = \dfrac{-3}{4}$    Cross-multiply.

$4x = (-3) \times 5$

$4x = -15$

$x = \dfrac{-15}{4}$ or $-\dfrac{15}{4}$

**33. (B) $(2, -1)$**

Substitute 2 for $x$ and $-1$ for $y$.

$\quad\quad 4x - y = 9$

$4 \times 2 - (-1) = 9$

$\quad\quad 8 + 1 = 9$    *true*

**34. (D) $y - 2x = 2$**

The $x$-intercept is $-1$ and the $y$-intercept is 2. In intercept form, the equation of the line is

$$\dfrac{x}{-1} + \dfrac{y}{2} = 1 \quad \text{Multiply both sides by 2.}$$

$-2x + y = 2$

or

$y - 2x = 2$

**35. (C) $-32$**

Substitute $-1$ for $a$ and 2 for $b$. Then

$c = 8ab^2 = 8(-1)2^2 = (-8) \times 4 = -32$

**36.** (C) $\dfrac{q+5}{4}$

$$4p - q = 5$$
$$4p - q + q = 5 + q$$
$$4p = q + 5$$
$$p = \dfrac{q+5}{4}$$

**37.** (D) 525

$$\dfrac{\text{hits}}{\text{at bats}} = \dfrac{\text{hits}}{\text{at bats}}$$

$$\dfrac{2 \text{ hits}}{7 \text{ at-bats}} = \dfrac{150 \text{ hits}}{x \text{ at-bats}}$$

$$\dfrac{2}{7} = \dfrac{150}{x} \quad \text{Cross-multiply.}$$

$$2x = 7 \times 150$$
$$2x = 1050$$
$$x = 525$$

**38.** (D) $\sqrt{17}$

$$a^2 + b^2 = x^2$$
$$4^2 + 1^2 = x^2$$
$$16 + 1 = x^2$$
$$17 = x^2$$
$$\sqrt{17} = x$$

**39.** (B) $x = 1$ and $y = 2$

$$3x + 2y = 7$$
$$\underline{3x - \phantom{2}y = 1} \quad \text{Subtract.}$$
$$3y = 6$$
$$y = 2$$

Substitute 2 for $y$ in the second equation.

$$3x - 2 = 1$$
$$3x = 3$$
$$x = 1$$

Thus $x = 1$ and $y = 2$

**40.** (B) 3000

6000 games at $5 per game costs
$30,000. The general overhead cost of
$15,000 brings the total costs to
$45,000. He receives $15 per game. To
break even, he must sell

$$\dfrac{\$45,000}{\$15} = 3000 \text{ (games)}$$

## Sample Final Exam B (page 271)

**1.** (C) 60,405

**2.** (C) 319

$$\begin{array}{r} 319 \\ 19\overline{)6061} \\ \underline{57} \\ 36 \\ \underline{19} \\ 171 \\ \underline{171} \end{array}$$

**3.** (D) 5 hours 15 minutes

$$\begin{array}{r} 9 \text{ hours} \quad 66 \text{ minutes} \\ \cancel{10 \text{ hours}} \quad \cancel{6 \text{ minutes}} \\ - \phantom{0}4 \text{ hours } 51 \text{ minutes} \\ \hline 5 \text{ hours } 15 \text{ minutes} \end{array}$$

**4.** (E) $\dfrac{29}{40}$

The *lcd* is 40.

$$\dfrac{1}{8} = \dfrac{5}{40} \qquad \dfrac{3}{5} = \dfrac{24}{40}$$

$$\dfrac{1}{8} + \dfrac{3}{5} = \dfrac{5}{40} + \dfrac{24}{40} = \dfrac{29}{40}$$

**5.** (C) 40

$$\dfrac{\$6.00}{\$.15} = \dfrac{600}{15} = 40$$

**6.** (D) 4

$$2\dfrac{3}{4}$$
$$\underline{1\dfrac{1}{4}}$$
$$3\dfrac{4}{4} = 3 + \dfrac{4}{4} = 3 + 1 = 4$$

**7.** (C) 1.11

$$\begin{array}{r} 3.09 \\ -1.98 \\ \hline 1.11 \end{array}$$

**8.** (E) $53

$$\begin{aligned} \text{PROFIT} &= \text{SALES} - \text{COSTS} \\ &= (70 \times \$1.50) - (80 \times \$.65) \\ &= \$105.00 - \$52.00 \\ &= \$53 \end{aligned}$$

**9.** (C) .72   9% of $8 = .09 \times 8 = .72$

**10.** (B) $24

A 25% reduction on a $32 sweater means
$.25 \times \$32$, which equals $8.

$$\begin{array}{rl} \$32 & \longleftarrow \text{ normal price} \\ - \phantom{0}8 & \longleftarrow \text{ reduction} \\ \hline \$24 & \longleftarrow \text{ new price} \end{array}$$

**11.** (A) $4.20

$$\dfrac{\text{cost}}{\text{ounces}} = \dfrac{\text{cost}}{\text{ounces}}$$

$$\dfrac{\$2.80}{8 \text{ ounces}} = \dfrac{\$x}{12 \text{ ounces}}$$

$$\dfrac{2.80}{8} = \dfrac{x}{12} \quad \text{Cross-multiply.}$$

$$12 \times 2.80 = 8x$$
$$33.60 = 8x$$
$$4.20 = x$$

The cost of a 12-ounce slice is $4.20.

**12.** (D) 81

$$\dfrac{72 + 85 + 80 + 87}{4} = \dfrac{324}{4} = 81$$

**13. (B)** 2

$$4\frac{1}{2} \div 2\frac{1}{4} = \frac{9}{2} \div \frac{9}{4} = \frac{\overset{1}{\cancel{9}}}{\underset{1}{\cancel{2}}} \times \frac{\overset{2}{\cancel{4}}}{\underset{1}{\cancel{9}}} = 2$$

**14. (B)** $\frac{1}{10}$     **15. (C)** 2.009     **16. (D)** 20.118

$$\begin{array}{r} 5.08 \\ 3.038 \\ \underline{12.} \\ 20.118 \end{array}$$

**17. (B)** $90.10

6% of $85 = .06 \times \$85 = \$5.10

$$\begin{array}{ll} \$85 & \longleftarrow \text{price} \\ + \quad 5.10 & \longleftarrow \text{sales tax} \\ \hline \$90.10 & \longleftarrow \text{total price} \end{array}$$

**18. (B)** $154

$$\frac{\text{earnings}}{\text{hours}} = \frac{\text{earnings}}{\text{hours}}$$

$$\frac{\$44}{8 \text{ hours}} = \frac{\$x}{28 \text{ hours}}$$

$$\frac{44}{8} = \frac{x}{28}$$

$44 \times 28 = 8x$

$1232 = 8x$

$154 = x$

He earns $154 for 28 hours of work.

**19. (C)** $45

Cost = Area $\times$ Cost per yd.$^2$

$$= (6 \text{ yd.} \times 5 \text{ yd.}) \times \frac{\$1.50}{\text{yd.}^2}$$

$$= 30 \overset{1}{\cancel{\text{yd.}^2}} \times \frac{\$1.50}{\underset{1}{\cancel{\text{yd.}^2}}}$$

$$= \$45.00$$

**20. (D)** 48

$$\begin{array}{l} 10 \longleftarrow \text{January} \\ 13 \longleftarrow \text{February} \\ 10 \longleftarrow \text{March} \\ \underline{15} \longleftarrow \text{April} \\ 48 \end{array}$$

**21. (C)** $4a^2 + 2a - 1$

$$\begin{array}{r} 4a^2 + \quad a - 3 \\ + \qquad\quad a + 2 \\ \hline 4a^2 + 2a - 1 \end{array}$$

**22. (D)** $-30a^3 b^4$

$$(-5ab) \cdot (6a^2 b^3) = \big((-5) \times 6\big)(a \cdot a^2)(b \cdot b^3)$$
$$= -30a^{1+2} b^{1+3}$$
$$= -30a^3 b^4$$

**23. (C)** $-6x^2$

$$2x^2 y - 2x^2(y + 3) = 2x^2 y - 2x^2 \cdot y - 2x^2 \cdot 3 = -6x^2$$

**24. (D)** $a + 9$

$$P - Q = P + (-Q)$$

$$\begin{array}{ll} P: & 5a + 4 \\ Q: & 4a - 5 \\ -Q: & -4a + 5 \end{array}$$

$$\begin{array}{r} 5a + 4 \\ -4a + 5 \\ \hline a + 9 \end{array}$$

**25. (D)** $2a - 6$

$$\frac{4a^2 - 12a}{2a} = \frac{4a^2}{2a} - \frac{12a}{2a} = 2a - 6$$

**26. (E)** $27x^{12} y^6$

$$(3x^4 y^2)^3 = 3^3 (x^4)^3 (y^2)^3$$
$$= 27x^{4 \times 3} y^{2 \times 3}$$
$$= 27x^{12} y^6$$

**27. (D)** $150

$$\underbrace{\text{price}}_{x} \times \underbrace{\text{tax rate}}_{.04} = \underbrace{\text{tax}}_{\$6}$$

$$4x = \$600$$
$$x = \$150$$

**28. (D)** $9x + 7y$

One pair of gloves costs 9 dollars.
$x$ pairs of gloves cost $x \cdot 9$ dollars or $9x$ dollars.
One scarf costs 7 dollars.
$y$ scarves cost $y \cdot 7$ dollars or $7y$ dollars.
The total cost, in dollars, is $9x + 7y$.

**29. (E)** $5xy(5x - 9)$

**30. (B)** 1

$$(-2)(-5) - (-3)^2 = 10 - 9 = 1$$

**31. (C)** 5

$$\frac{x + 3}{2} = \frac{3x - 3}{2} \quad \text{Cross-multiply.}$$

$$3(x + 3) = 2(3x - 3)$$
$$3x + 9 = 6x - 6$$
$$9 = 3x - 6$$
$$15 = 3x$$
$$5 = x$$

**32. (D)** $(8, 2)$

Substitute 8 for $x$ and 2 for $y$.

$$3x - 2y = 20$$
$$3 \times 8 - 2 \times 2 = 20$$
$$24 - 4 = 20 \quad true$$

**33. (B)** −21

Substitute 4 for $a$ and −1 for $b$ in the expression $5ab - b^2$.

$$5 \times 4 \times (-1) - (-1)^2 = -20 - 1 = -21$$

**34. (D)** $2x + 3y = 6$

The $x$-intercept is 3 and the $y$-intercept is 2. In intercept form, the equation of the line is

$$\frac{x}{3} + \frac{y}{2} = 1$$

or

$$2x + 3y = 6$$

**35. (C)** $\dfrac{24}{5}$

$$5x - 2 = 22$$
$$5x - 2 + 2 = 22 + 2$$
$$5x = 24$$
$$x = \frac{24}{5}$$

**36. (D)** $\dfrac{b+c}{4}$

$$4a - b = c$$
$$4a - b + b = c + b$$
$$4a = b + c$$
$$a = \frac{b+c}{4}$$

**37. (A)** $x = 2$, $y = 1$

$$2x + y = 5$$
$$5x - y = 9 \quad \text{Add.}$$
$$7x = 14$$
$$x = 2$$

Substitute 2 for $x$ in the first equation.

$$2 \times 2 + y = 5$$
$$4 + y = 5$$
$$y = 1$$

Thus $x = 2$ and $y = 1$

**38. (C)** $10\dfrac{1}{2}$ inches

$$\frac{\text{width}}{\text{length}} = \frac{\text{width}}{\text{length}}$$

$$\frac{5 \text{ inches}}{7 \text{ inches}} = \frac{7\frac{1}{2} \text{ inches}}{x \text{ inches}}$$

$$\frac{5}{7} = \frac{7.5}{x} \quad \text{Cross-multiply.}$$

$$5x = 7 \times 7.5$$
$$5x = 52.5$$
$$\frac{5x}{5} = \frac{52.5}{5}$$
$$x = 10.5 \text{ or } 10\frac{1}{2}$$

The length of the enlargement is $10\frac{1}{2}$ inches.

**39. (D)** 8

$$\sqrt{10^2 - 6^2} = \sqrt{100 - 36} = \sqrt{64} = 8$$

**40. (D)** 302

Let $x$ be the number of tickets sold.

Profit = Sales − Expenses
$$\$1156 = x \cdot \$8 - \$1260$$
$$\$2416 = \$8x$$
$$2416 = 8x$$
$$302 = x$$

## Sample Final Exam C (page 275)

**1. (C)** 80,000.4    **2. (B)** 504    **3. (B)** 11.5   $\dfrac{10 + 11 + 12 + 13}{4} = \dfrac{46}{4} = 11.5$

**4. (D)** 3 pounds 15 ounces

$$\begin{array}{r} 9 \text{ pounds} \quad 27 \text{ ounces} \\ \cancel{10 \text{ pounds } 11 \text{ ounces}} \\ - \ 6 \text{ pounds } 12 \text{ ounces} \\ \hline 3 \text{ pounds } 15 \text{ ounces} \end{array}$$

**5. (D)** $-\dfrac{1}{40}$

The *lcd* is 40.    $\dfrac{3}{8} = \dfrac{15}{40}$,    $\dfrac{2}{5} = \dfrac{16}{40}$

$$\frac{3}{8} - \frac{2}{5} = \frac{15}{40} - \frac{16}{40} = -\frac{1}{40}$$

**6. (B)** $\dfrac{2}{3}$

$$\overset{1}{\underset{1}{\cancel{\frac{2}{5}}}} \times \overset{2}{\underset{3}{\cancel{\frac{10}{9}}}} = \frac{2}{3}$$

**7. (A)** $\dfrac{2}{3}$    $\dfrac{3}{10} \div \dfrac{9}{20} = \overset{1}{\underset{1}{\cancel{\frac{3}{10}}}} \times \overset{2}{\underset{3}{\cancel{\frac{20}{9}}}} = \dfrac{2}{3}$

**8. (B)** 5

$$\begin{array}{r} 3\frac{3}{4} \\ + 1\frac{1}{4} \\ \hline 4\frac{4}{4} = 4 + \frac{4}{4} = 4 + 1 = 5 \end{array}$$

**9. (D)** $\dfrac{5}{12}$

**10. (D)** .0289    **11. (B)** 22.812

$$\begin{array}{r} 6.003 \\ 5.809 \\ 11. \\ \hline 22.812 \end{array}$$

**12.** (D) $20

  1 gallon costs $1.25.
  16 gallons cost $16 \times \$1.25$ or $20.00

**14.** (D) .31

$$13\overline{)4.000}$$ with quotient .307

  $\dfrac{39}{100}$
  $\dfrac{91}{\phantom{9}}$

  The third decimal digit is 5 or more. To the nearest hundredth,

  $\dfrac{4}{1.3} \approx .31$

**13.** (C) $1.67

  1 pound of potatoes costs $.22.
  5 pounds of potatoes cost $5 \times \$.22$ or $1.10.
  1 pound of onions costs $.19.
  3 pounds of onions cost $3 \times \$.19$ or $.57.

  $\begin{array}{r} \$1.10 \\ +\ \ .57 \\ \hline \$1.67 \end{array}$

**15.** (E) 64

  Let $x$ be the number.

  48 is 75% of a certain number.
  $48 = .75 \times x$

  Multiply both sides by 100.

  $4800 = 75x$
  $\dfrac{4800}{75} = x$
  $64 = x$

**16.** (B) 36 tons

  $30\% \times 120$ tons $= .30 \times 120$ tons
  $= 36.00$ tons

**17.** (C) $45

  A 25% discount on $60 means .25 × $60 or $15.

  $\begin{array}{r} \$60 \leftarrow \text{original price} \\ -\ \ 15 \leftarrow \text{discount} \\ \hline \$45 \leftarrow \text{new price} \end{array}$

**18.** (E) $768

  Cost = Area × Cost per ft.²
  $= (16 \text{ ft.} \times 12 \text{ ft.}) \times \dfrac{\$4}{\text{ft.}^2}$
  $= 192 \text{ ft.}^2 \times \dfrac{\$4}{\text{ft.}^2} = \$768$

**19.** (D) 21

  $(-2-1)^2 - 4(-3) = (-3)^2 - (-12) = 9 + 12 = 21$

**20.** (A) 500

  $\begin{array}{r} 4000 \leftarrow \text{1979 enrollment} \\ -\ 3500 \leftarrow \text{1978 enrollment} \\ \hline 500 \leftarrow \text{increase} \end{array}$

**21.** (C) $4x - 5$

  5 less than 4 times the number
  $-5 + \quad 4 \quad \cdot \quad x$  or $4x - 5$

**22.** (D) $-x^2 + 3x - 7$

  $P - Q = P + (-Q)$
  $P:\ x^2 - 6$
  $Q:\ 2x^2 - 3x + 1$
  $-Q:\ -2x^2 + 3x - 1$

  $\begin{array}{r} x^2 \qquad -6 \\ -2x^2 + 3x - 1 \\ \hline -x^2 + 3x - 7 \end{array}$

**23.** (B) $4a^3b - 8a^2b^3$

  $4ab(a^2 - 2ab^2) = 4ab \cdot a^2 - 4ab \cdot 2ab^2$
  $= 4a^{1+2}b - (4 \times 2)a^{1+1}b^{1+2}$
  $= 4a^3b - 8a^2b^3$

**24.** (A) $12x^8y^4$

  $3x^2(2x^3y^2)^2 = 3x^2(2^2(x^3)^2(y^2)^2)$
  $= 3x^2(4x^{3\times2}y^{2\times2})$
  $= 3x^2(4x^6y^4)$
  $= (3 \times 4)x^{2+6}y^4$
  $= 12x^8y^4$

**25.** (B) $3x^3 - 5x$

  $\dfrac{15x^4 - 25x^2}{5x} = \dfrac{15x^4}{5x} - \dfrac{25x^2}{5x}$
  $= 3x^3 - 5x$

**26.** (C) $3ab(4a^2 - 3)$

**27.** (B) 8

  $4x - 7 = 2x + 9$
  $2x - 7 = 9$
  $2x = 16$
  $x = 8$

**28.** (E) 5

  $\dfrac{t+3}{4} = \dfrac{t+1}{3}$  Cross-multiply.
  $3(t+3) = 4(t+1)$
  $3t + 9 = 4t + 4$
  $9 = t + 4$
  $5 = t$

**29.** (A) $4 - 3a + 2b$

  $\dfrac{3a - 2b + c}{4} = 1$
  $3a - 2b + c = 4$
  $3a - 3a - 2b + 2b + c = 4 - 3a + 2b$
  $c = 4 - 3a + 2b$

**30. (D)** $(6, 4)$

Substitute 6 for $x$ and 4 for $y$.
$$4 \times 6 - 3 \times 4 = 12$$
$$24 - 12 = 12 \quad true$$

**31. (B)** $-2$

Substitute 0 for $x$ and $b$ for $y$.
$$5 \times 0 - 2b = 4$$
$$0 - 2b = 4$$
$$-2b = 4$$
$$b = -2$$

**32. (C)** $3x - 2y = 6$

The $x$-intercept is 2 and the $y$-intercept is $-3$. In intercept form, the equation of the line is

$$\frac{x}{2} + \frac{y}{-3} = 1 \quad \text{Multiply both sides by 6.}$$

$$3x - 2y = 6$$

**33. (D)** $x = 3$, $y = 7$

Multiply both sides of $5x - y = 8$ by 2.

$$10x - 2y = 16$$
$$\underline{3x - 2y = -5} \quad \text{Subtract.}$$
$$7x \quad\quad = 21$$
$$x \quad\quad = 3$$

Substitute 3 for $x$ in the equation $5x - y = 8$.
$$5 \times 3 - y = 8$$
$$15 - y = 8$$
$$15 = 8 + y$$
$$7 = y$$
Thus $x = 3$ and $y = 7$

**34. (B)** 24

Substitute $-2$ for $p$ and 3 for $q$.
$$3(-2)^2 - 2(-2)3 = 3 \times 4 - (-12)$$
$$= 12 + 12$$
$$= 24$$

**35. (D)** $\frac{3}{8}$

Substitute 3 for $x$.
$$\frac{3}{2^3} = \frac{3}{8}$$

**36. (D)** $5n + 25q + 3$

The value of 1 nickel is 5 cents.
The value of $n$ nickels is $n \times 5$ cents or $5n$ cents.
The value of 1 quarter is 25 cents.
The value of $q$ quarters is $q \times 25$ cents or $25q$ cents.
The value of 3 pennies is 3 cents.
The total value, in cents, is $5n + 25q + 3$.

**37. (C)** 13.5 pounds

$$\frac{\text{number of pounds}}{\text{number of rats}} = \frac{\text{number of pounds}}{\text{number of rats}}$$

$$\frac{9 \text{ pounds}}{50 \text{ rats}} = \frac{x \text{ pounds}}{75 \text{ rats}}$$

$$\frac{9}{50} = \frac{x}{75} \quad \text{Cross-multiply.}$$
$$9 \times 75 = 50x$$
$$675 = 50x$$
$$13.5 = x$$

**38. (D)** 12

$$\frac{x}{3} - 2 = \frac{x-2}{5} \quad \begin{array}{l}\text{Multiply both} \\ \text{sides by 15.}\end{array}$$

$$15\left(\frac{x}{3} - 2\right) = \overset{3}{\cancel{15}} \cdot \frac{x-2}{\underset{1}{\cancel{5}}}$$

$$\frac{15x}{3} - 30 = 3(x - 2)$$
$$5x - 30 = 3x - 6$$
$$2x - 30 = -6$$
$$2x = 24$$
$$x = 12$$

**39. (E)** $\sqrt{15}$

$$a^2 + b^2 = c^2$$
$$x^2 + 1^2 = 4^2$$
$$x^2 + 1 = 16$$
$$x^2 = 15$$
$$x = \sqrt{15}$$

**40. (C)** 30

25% of $60 = .25 \times 60 = 15$
Thus 15 of the guidance counsellors are women.
If $x$ additional women are hired, then
$\dfrac{\text{number of women counsellors}}{\text{total number of counsellors}}$ will be 50%.

$$\frac{15 + x}{60 + x} = \frac{1}{2} \quad \text{Cross-multiply.}$$
$$2(15 + x) = 60 + x$$
$$30 + 2x = 60 + x$$
$$30 + x = 60$$
$$x = 30$$

# Index

www.ingramcontent.com/pod-product-compliance
Lightning Source LLC
Chambersburg PA
CBHW061340210326
41598CB00035B/5835